WORKING THE SAHEL

Since the 1970s the Sahel has been portrayed as a place of actual or threatening disaster, where people suffer from and in turn cause environmental degradation and desertification. This book goes beyond these stereotypes to describe the ways in which farming households in the Sahelian region organise themselves economically to secure their livelihoods.

Drawing on four years of field research with farmers in the Sahelian region of north-east Nigeria, and building on work with these communities over several decades, *Working the Sahel* looks at how people in the semi-arid conditions of the Sahel cope with their harsh environment, and in particular examines the ways in which they organise their labour to manage fields, crops and other resources.

Working the Sahel analyses the diversity, flexibility and adaptability that are the critical attributes of successful Sahelian systems of resource management. Reporting on studies of four village communities and their natural environments, it examines the ultimate cause of much environmental variability in the Sahel: the rainfall and its characteristics. The authors look at how farmers manage biological resources, crop and non-crop biodiversity and soil fertility, and transform the landscape through agricultural intensification. They show how gender, age and the division of labour interact, and how women and children make essential contributions to household viability. The authors conclude with an examination of differentiation between households, and try to define poverty in a rural Sahelian context, as well as placing issues in a broader policy context.

Working the Sahel presents important new evidence to indicate that the 'crisis' of degradation in the Sahel can be contained, and indeed is being contained in some areas, through the work of rural communities themselves.

Michael Mortimore is a Senior Research Associate in the Department of Geography, University of Cambridge and **Bill Adams** is a Lecturer in the Department of Geography, University of Cambridge.

ROUTLEDGE RESEARCH GLOBAL ENVIRONMENTAL CHANGE SERIES

WORKING THE SAHEL

Environment and society in northern Nigeria

Michael Mortimore and
William M. Adams

Global Environmental Change Programme

London and New York

First published 1999
by Routledge
11 New Fetter Lane, London EC4P 4EE

Simultaneously published in the USA and Canada
by Routledge
29 West 35th Street, New York, NY 10001

Routledge is an imprint of the Taylor & Francis Group

Typeset in Galliard by
J&L Composition Ltd, Filey, North Yorkshire
Printed and bound in Great Britain by
TJ International Ltd, Padstow, Cornwall

British Library Cataloguing in Publication Data
A catalogue record for this book is available
from the British Library

Library of Congress Cataloging in Publication Data
Mortimore, Michael, 1937–
Working the Sahel: environment and society in northern Nigeria/
Michael Mortimore and William M. Adams.
(Global environmental change series)
Includes bibliographical references and index.
1. Human ecology – Sahel.   2. Desertification – Sahel.   3. Desert
ecology – Sahel.   4. Sahel – Social conditions.   5. Sahel –
Environmental conditions.   I. Adams, W.M. (William Mark), 1955– .
II. Title.   III. Series.
GF740.M67   1999
306.3′49′0966–dc21   98–34516

ISBN 0–415–14096–X (hbk)

The farmer waits for the precious fruit of the earth, being patient over it until it receives the early and the late rains.

James 5: 7

CONTENTS

ILLUSTRATIONS

Plates

TABLES

ACKNOWLEDGEMENTS

This book draws on research carried out collaboratively between the departments of geography at Bayero University Kano and the University of Cambridge between 1992 and 1997. We are grateful to the Nigerian Universities Commission, the Vice-Chancellor of Bayero University, and the Head of the Department of Geography for this long and fruitful association. The work was funded by two research projects held jointly by the authors with Professor Michael Chisholm (University of Cambridge), working with the collaboration of Dr J. Afolabi Falola in Kano. Without their tireless support the work could not have been completed. We are particularly grateful for Michael's determination to see the research funded in the first instance and his constant support throughout, and for Afolabi's enduring kindness, commitment and efficiency.

The field research, however, draws on the work of a wider range of people, chief among whom are Malam Ahmed Ibrahim, Malam Salisu Mohammed and Malam Maharazu Yusuf, now Assistant Lecturers at Bayero University, and Alhaji Maigana Chiroma, Lecturer in Agricultural Sciences at Ramat Polytechnic, Maiduguri. Over the life of the project, their contribution as researchers grew steadily, and theirs was the often exhausting work of maintaining a field programme through many difficulties including fuel shortages and sickness. The insights we have gained owe much to their persistence and diligence. In the first two years of our work we benefited from an indispensable contribution by Malam Aminu Shehu. We would also like to thank those who made specialist contributions to our project: Professor Emmanuel A. Olofin, Luka F. Buba, Dr E.U. Essiet and Dr Sebastian Patrick (of Bayero University), Dr David Ogbonna (of Ahmadu Bello University) and Malam Usman Geidam (of Kano). The Director of the British Council, Kano, Rajiv Bhendre, kindly put his office at our disposal to facilitate communications between Cambridge and Kano. Many others helped us in other ways, especially Malam Adamu Gombe. The Chairmen of Gezawa, Birniwa, Yusufari and Geidam Local Government Authorities gave their approval and support to our field investigations.

The research was funded first under the ESRC's Global Environmental Change Programme (*Agropastoral Adaptation to Environmental Change in Northern Nigeria*, Project No. L320253001), and later by the Natural

Resources Systems Programme (Semi-Arid Production Systems) of the Department for International Development (DFID), formerly the Overseas Development Administration (*Soils, Cultivars and Livelihoods in Northeastern Nigeria*, Project No. R 6051). DFID can accept no responsibility for any information provided or views expressed. We are very grateful to David Jackson (NRI) and Chris Lewcock (NRIL) for their patience and support, and for their inputs at critical times to the research.

We have also gained greatly from interaction with other researchers, particularly Frances Harris, with whom we have worked very closely in two of our survey villages, and Beryl Turner, whose expertise in land use change mapping has been generously contributed. We would also like to thank our collaborators, Richard Carter and John Page, for their work on groundwater and climate change. We have had a great deal of help with data management, particularly in Kevin Kimmage's heroic creation and manipulation of the *Access* database, and also from John Thompson, Keith Turner and Simon Benger. Owen Tucker, Mike Young and Philip Stickler drew many of the diagrams, and Beryl Turner contributed Figures 3.1–3.4 and 5.1. Over the life of the projects we have gained from discussions with many people, particularly John Murton, Simon Batterbury and Mary Tiffen, and we would like to thank all those who have participated in the various seminars and workshops where we have had opportunities to discuss the work and its implications. Michael Chisholm and Beryl Turner gave generously of their time to read a draft manuscript.

While a book of this kind draws most immediately on specific research activities, it also draws on much longer and deeper roots. This work is the latest episode in Michael Mortimore's long engagement with northern Nigeria, and an association with some of the villages where the research was carried out that spreads back more than a quarter of a century. We would like to offer our sincere thanks to the Village Heads and the people of those four villages for their patience and hospitality. For each of us their generosity with time and resources remains salutary and quite wonderful. In return, we offer the hope that the research reports their understanding of their landscapes and livelihoods honestly and accurately, and perhaps that it will help others to understand their situation better. We have limited faith in formal 'uptake pathways', let alone in the capacity of researchers to ignite positive developmental change through their writings. The most effective pathway for innovations in northern Nigeria remains, as always, the ideas and hard work of local people, and we are honoured to have been allowed to glimpse them at work.

Many other friends in Nigeria and the UK have helped us in the research and in writing this book. We would like to thank those who supported us, particularly in the aftermath of the bizarre and unreasonable events of October 1996. We would like above all to thank Julia and Franc, who have endured our various moments of restless energy and desperation over the last five years.

This book is the tip of an iceberg, both in terms of the vast amount of research done by all our collaborators, and the huge amount more that remains to be understood about people and nature in the Sahel. The drylands of West Africa are not homogeneous, but wonderfully diverse; not a place, but a kaleidoscope of places, each changing very fast, each the home of people who are active and reflexive agents of change. We hope that our discussions will be seen as starting, and not seeking to close down, debate about those places, and the choices that those people face about the future.

Michael Mortimore and William M. Adams
Milborne Port and Cambridge

1

INTRODUCTION

Mai ha'kuri ya kan dafa dutse ya sha romonsa
(A patient person will cook a stone and drink its broth)
(Hausa proverb, Abrahams, 1949)

The Sahel: beyond crisis

The *Sahel crisis* was discovered by the outside world during the great Sahel drought of 1969–1974. It has been with us, in one form or another, ever since. This was the first major African famine to dominate the media headlines in the North, and it provoked an anguish, both in scientific and in aid circles – whether bilateral or multilateral donors, or charities – which was itself new, creating a sharp discontinuity with the optimistic outlook that had tended to prevail during the 1960s, the post-independence era.

Hitherto, the impoverished but productive societies who inhabited the Sahel of West Africa, using mainly hand or simple animal-powered technologies, had produced enormous exports of groundnuts and cotton for the world markets. They had contributed large shares to the GNPs of their countries. Indeed, without them several national economies would not have been viable. In addition, the Sahel produced food grains (cereals, principally millet and sorghum, and legumes, principally cowpeas) and meat for growing numbers of urban consumers. In the coastal belt, where foreign investment was concentrated, the port-capitals spearheaded national drives to achieve economic development by import-substituting industrialisation. They needed labour, and this too the Sahel could supply, sometimes in the form of permanent migration, but more usually by means of short-term circulation of dry-season labour. The cocoa plantations of the forest, and the mining areas (for example, the gold fields of southern Ghana and the tin fields of central Nigeria) added to the employment opportunities available under the colonial economic systems.

Thus both regional and global markets were integral to the Sahelian economies, and while the intensity of these market relationships reached a peak in the 1950s, shortly before independence, they had deep historical roots in old migratory and exchange systems. The idea of a Sahel buried in isolation from the rest of the world and preoccupied exclusively with subsistence is an illusion. Nevertheless, Sahelian households, generally speaking, continue even today to set the highest priority on subsistence, using mainly family labour to fill their granaries, if possible, during each shortlived and fickle rainy season. This is the

Plate 1.1 Preparing groundnut ridges in the early rains on the annually cultivated fields of the Kano Close-Settled Zone near Tumbau.

basis of economic security and the precondition for social reproduction, even as markets penetrate everyday aspects of life, monetising not only material goods but social transactions such as the exchange of labour, and natural resources such as land and fodder.

Then came drought: up to seven years of rainfall below the average, culminating in regional crop failures in 1972 and 1973, the decimation of livestock holdings, the abandonment of settlements, the displacement of pastoral groups, accelerated migrations, and changes in natural habitats, such as the succession of annual grasses in place of perennials, and the reactivation of sand-dunes previously under vegetation. Up to six million people were said to be threatened with starvation. Productive systems were said to have 'collapsed' or 'broken down'. It was 'the quintessence of a major environmental emergency' (Raynaut *et al.*, 1997: 1). Compulsory resettlement was advocated, as it was believed that whole ecozones were incapable of supporting such populations again; indeed, they were diagnosed, in a post hoc scientific rationalisation, to have already exceeded their 'carrying capacity' several years before. Proposals were brought forward for planting a barrier of millions of trees along the southern boundary of the Sahara to stop its further advance; or to zone land use into areas suitable for livestock breeding, stock fattening and crop production, which would have had draconian implications for rural Sahelians.

The Sahel Drought brought an international identity to a biogeographical

2

Plate 1.2 Product of smallholders' labour: half a million tons of groundnuts await export in the Kano 'pyramids' of 1967.

zone, seemingly marked out from the rest of Africa, threatened by natural disaster and brimming with human misery. In a growing literature on environmental crises, the Sahel had a privileged, if unenviable place. The grounds of its distinctiveness shifted, however, from those of ecology or rainfall to that of a geopolitical entity, a region whose chronic poverty and dependency on food aid came to define the terms of its engagement with the rest of the world (Somerville, 1986). To foster the flow of aid, coordinate the requests of its member countries and maintain the political links that were necessary, the CILSS (Comité permanent Inter-état de Lutte contre la Sécheresse dans le Sahel) and the Club du Sahel were set up in West Africa and Europe respectively. Increasingly, the membership of the CILSS (which was composed of the formerly French colonial territories now Senegal, Mauritania, Mali, Burkina Faso, Niger and Chad) came to define public perceptions of the region, and even those of the international scientific community. In this book, we use a biogeographical definition, including a substantial portion of northern Nigeria and smaller portions of north Cameroon and Ghana.

Scientists settled down to analyse and debate the causes of the disaster, quickly exposing a divergence between, on the one hand, a view which saw in it the consequences of rainfall failure – and, as became more clear with the passage of time, a quasi-permanent decline in average rainfall – and, on the other, a view which saw its roots in colonial export agriculture and structural change. The first view led naturally to a negative evaluation of what were seen to be the consequences of uncontrolled population growth: the 'inappropriate' land use practices of overcultivation, overgrazing and deforestation. The second view placed the blame for these forms of degradation squarely on export agriculture which, driven by monetary taxation, coercive production and the monetisation of the economy, weakened self-sufficiency and economic auton-omy in rural communities and made them vulnerable to depressed prices. The first view was given expression in the Plan of Action to Combat Desertification which was approved by the United Nations Conference on Desertification in 1977, proposing mainly technical interventions which, by and large, had a disappointingly small impact on incomes and welfare in the Sahel (United Nations, 1977; Sinclair and Fryxell, 1985; Swift, 1996; Warren, 1996). The second view, inspired initially by Marxist interpretations of economic change (Copans, 1983; Watts, 1983, 1984), also failed to produce solutions. Both are now generally agreed to have placed too much reliance on simplistic and generalised hypotheses of ecological change. By contrast, more recent research shows hitherto unsuspected ambivalences in processes of ecological change, and has begun to uncover the enormous diversity and complexity of Sahelian production systems and their interactions.

Another major drought cycle occurred in the early 1980s, renewed uncer-tainty in the rainfall returned (with El Niño) in 1996–1997, and there is general agreement that mean rainfall has declined by as much as one-third in some areas of the Sahel since the 1960s (Hulme, 1992). Many failures among development projects and programmes meanwhile forced a reevaluation of perceptions of degradation (Nelson, 1988) and of orthodox technical approaches to containing the 'Sahel problem'. A recent review of activities managed by the World Bank, for example, led to the conclusion that

> the traditional dryland management projects supported by the Bank and other donors have had mixed success, and call for a 'new' approach signified by a more profound understanding of the rationality of tradi-tional [sic] practices, greater reliance on local community institutions, more open-ended and flexible programs, greater participation of all stakeholders in shaping the interventions, and the provision of an enabling environment that provides market-driven incentives for change.
>
> (Bojö and Chee, 1997: v)

Yet notwithstanding such failures, the Sahel has produced some surprises. Population growth rates have been maintained, if more slowly than in the humid zone, nevertheless at about 2 per cent per year (IUCN, 1989). Food sufficiency in most years, and in most countries, has recovered. Between 1980 and 1990, food imports fell from 21 to 14 per cent of consumption in the region, and food production increased in all the countries of the CILSS except for the Gambia and Cape Verde (Cour, 1994); though following poor rainfall in 1995–1996, the situation has deteriorated, at least in Niger (Club du Sahel, 1998). Some areas have dramatically increased their output of food crops for internal markets. Lowlands where irrigation, or flood-recession farming, is possible have received increased investment from small farmers (Kimmage, 1991a; Adams, 1992; Kimmage and Adams, 1992; Hollis *et al.*, 1993). Livestock holdings recovered quite quickly (in four to six years) after major periods of drought-induced mortality. Numbers have increased in some areas, such as southern Mali from 1982 to 1993 (Bade *et al.*, 1997: 4), and northern Nigeria (RIM, 1992), where there has been a significant shift from cattle to small ruminants (de Leeuw *et al.*, 1996). The density of livestock has been found, in aerial surveys, to correlate quite specifically with population density, contrary to expectations (Bourn and Wint, 1994). Notwithstanding rapid urbanisation (within as well as outside the Sahel), rural population densities have continued to increase, notably in the hinterlands of major cities, where there is now evidence of significant intensification in smallholder farming (Snrech *et al.*, 1994). This intensification, however, is not of the type promoted by the cotton corporations and agricultural development programmes, namely dependent on the use of inorganic fertilisers, but is dependent primarily on the use of organic manure and additional labour per hectare.

The geopolitical definition of the Sahel as a group of formerly French colonies is at variance with the ecological reality of a 'greater Sahel' which includes a vast and populous area in northern Nigeria (Chapter 1), where more ambivalent diagnoses, and scenarios for the future, have been offered. Indeed a 'present trends scenario' for the Sahel as a whole has been challenged (Giri, 1988). Dryland Africa as a whole, and the Sahel in particular, have been repeatedly portrayed in international literature as places where production systems have failed to achieve sustainability, where human over-use of nature has caused degradation, where population growth and drought threaten a Malthusian crisis, and where development interventions have themselves struggled and failed to yield sustained benefits (Commins *et al.*, 1986; Watts, 1989; Morgan and Solarz, 1994). In the Sahel, crises of environment, development intervention and governance seem to vie with each other in the severity of the threat they are claimed to offer to prospects for sustainable development (Adams, 1990). However, in fact both agricultural and pastoral systems have proved surprisingly resilient in the Sahel, and their productivity has been sustained to a remarkable degree.

The 'Sahel problem' remains a transnational issue, though institutional responses at the national level (erosion control, afforestation, agricultural and other development programmes) have dominated efforts to solve it. Whether at the local, national or international level, many interests besides those of farmers and stockraisers (bureaucracies, businesses, donors, researchers) are implicated in what are, in effect, prolonged environmental negotiations (Mortimore, 1993c). Nevertheless, 'desertification' was not included in the original agenda for the Rio Earth Summit (1992). Debates about sustainable development, there and at other international fora, have been extensive (Chattergee and Finger, 1994; McCormick, 1992). After the Earth Summit, the International Convention to Combat Desertification has revived official anxiety, and the preparation of national action plans is receiving donor assistance in several Sahelian countries. Thus far, they have had little impact on the attempts of Sahelian producers to achieve sustainable livelihoods and to maintain the sustainability of their environmental management, or on outsiders' understanding of their success (or failure) in doing so.

In the study of pastoral production in African arid environments, a revolution has meanwhile occurred in the way in which livestock management and grazing practice are understood. It is now widely agreed that in such variable environments, where rainfall and pasture resources are both temporally and spatially unpredictable, the rationale of indigenous herding is *opportunistic,* rather than being based on the minimal supporting capacities of pastures; and furthermore that this is the most efficient way of using such resources to support the multipurpose herds required for maintaining pastoral families (Sandford, 1983; Behnke and Scoones, 1991; Behnke *et al.*, 1993; Scoones, 1994). In other words, it is better to take advantage of good years by building up the herds through breeding, even though when bad years come, heavy losses must be sustained. It is not very useful, therefore, to speak of 'carrying capacities' of rangeland, as they vary from year to year. Neither is it helpful to perceive of ecosystems as *equilibrial* in nature when, in fact, they sustain enormous shocks from year to year (Ellis and Swift, 1988). Rather their ability to sustain themselves under variable rainfall (*disequilibrial* behaviour) testifies to their resilience.

In the era of the El Niño oscillation of the 1990s, with its worldwide repercussions on weather, disequilibrium is not a surprising idea. Holling (1973), who first defined the 'unstable but resilient' ecosystem, later showed that when the time-frame is lengthened, *any* ecosystem may be vulnerable to extreme or 'surprise' events, whether they emanate from outside or within the system, as 'accidents waiting to happen' (1987). Disequilibrium has entered debates on the management of natural forests and parks, such as Yellowstone.

For farmers, risk taking is also a way of life (Mortimore, 1998). Every year the seed is sown, and much back-wearying labour expended, for a probability of return that would be so low as to discourage many a farmer in a more temperate clime. Food security depends on accumulating reserves against harvest failure, and on swift adaptations to changing conditions. With hindsight, it is now apparent that the disaster that took the world by storm in 1972–1974 was not unprecedented, and that Sahelian farming systems contain in-built risk compensating mechanisms. Many of the surprises sprung by the Sahelians after 1974 are attributable to these resources. They include, of course, access to the employment and trading opportunities referred to above, both in Sahelian towns and outside the region. Today, in many areas of the Sahel, more people maintain a hold on livelihoods, at greater population and livestock densities, than before the drought of 1974.

These realities suggest that there is much to be learned from a better understanding of Sahelian production systems. The diversity of these systems, which is becoming more and more apparent, discourages generalisation of the kind which has misled in the past. Raynaut (1997) and his collaborators have provided a painstaking documentation of this diversity. We agree that a priority is to identify representative types of system, on the basis of case studies analysed in some depth. Such studies can achieve several things. They can provide an understanding of the natural complexity which African smallholders learn – by inherited, collective and individual experience – to manage. They show how natural resources are 'socially constructed' out of the knowledge and capabilities of indigenous peoples, using in an incremental way new technologies that are useful to them, rather than being brought into existence through the technical assessments of outsiders. If continued over a period of several years, and set in their context of longer term environmental change, case studies can put present practice into time perspective – a perspective which approaches that of rural families. They can also help focus attention on the capabilities (as well as the incapacities) of productive systems – their adaptability, flexibility and sustainability under present conditions of change (Mortimore, 1989; Adams and Mortimore, 1997).

Most intriguing of all, case studies provide a method for putting the hypothesis of a Sahel crisis to the test. For this is incumbent on us. For a quarter of a century, a diagnostic paradigm of environmental and developmental crisis has been accepted in scientific and policy debate about the Sahel with little question. It is now clear that it began its course on the basis of highly debatable (if not wrong) assumptions and predictions. Evidence for degradation there certainly is, but its interpretation is often ambiguous and its dimensions disputed (Warren, 1996). The processes involved have been oversimplified (Leach and Mearns, 1996b). Furthermore, its power has been linked to the scale of analysis. Degradational scenarios depend heavily (though not exclusively) on global or regional assessments or projections (Warren and Khogali, 1991). It is often a subject of comment that studies at the micro-scale tend to lead to less alarming perceptions of ecological change. While problems are acknowledged by smallholders, they are very often *not* the problems which environmental science wishes to highlight. What is the reason for such incongruities?

Poor families have to manage their constraints – of labour time and energy, of soil fertility, of livestock, of biodiversity, of livelihood options, and above all, of rainfall – if they are to survive (Netting, 1993; Mortimore, 1989, 1998). In their modes of management there is much wisdom, as is evidenced by the simple fact of their survival in such challenging circumstances (and in particular, during the past 25 years). There are also many failures, and the reasons for these need to be understood. We confess ourselves to be fascinated by the exercise of these skills, by people who have too often been dismissed as agents of 'indiscriminate' deforestation, nutrient mining or other environmental sins. We therefore sought to devise an empirical approach for uncovering some of this behaviour in small numbers of households in several widely different communities in the Nigerian Sahel. It is clear that the management of natural resources must deal effectively with both the variability of the rainfall and the ever-increasing density of the rural population (which translates into an intensifying scarcity of land, whether for cultivation or grazing). These forms of adaptation are therefore uppermost in our research design.

In the next chapter we shall sketch the conceptual basis of the present study, using a model of constraints and responses in which *diversity*, *flexibility* and *adaptability* are posited as critical attributes of successful Sahelian systems of resource management. A central place in this scheme is taken by labour, which, as it is related to historical population growth, tends to substitute land for labour as the most limiting factor. The method of this study is then described. We shall then introduce our four village communities and their natural environments (Chapter 3). In Chapter 4 we examine the ultimate cause of much environmental variability in the Sahel: the rainfall and its characteristics. Data from the four villages are used to show how farmers manage this variability. In Chapter 5, we investigate how farmers manage their biological resources, and in particular, crop and non-crop biodiversity, maintain soil fertility and transform the landscape. Chapter 6 is a systematic investigation of the evidence for agricultural intensification, in space and in time. In Chapter 7 we will examine the diversification of household income activity from crop production through livestock to off-farm activities and distant places. In Chapter 8 we show how gender, age and the division of labour interact, and how women and children make essential contributions to household viability. In Chapter 9 we look at the forms of differentiation between households, and try to define poverty in a rural Sahelian context. Finally, in Chapter 10, we place our findings in a broader policy context.

The Sahel

As we have suggested, the Sahel is a diverse, unpredictable and harsh environment in which to pursue a livelihood, if judged by living standards which westerners (and many other Africans) have come to take for granted. These characteristics also severely limit the scope of the economic and technical assistance which governments and other agencies can provide. It is imperative to understand as far as possible the indigenous experience, knowledge and skills whereby the natural resources are managed.

Sahelian ecosystems differ from those of humid and sub-humid Africa not only in the quantified variables such as rainfall and temperature but also in the intensity of the adaptive challenge which they pose for their human communities. In seeking an understanding of the day-to-day decisions made by small farmers and stockowners in the Sahel, the unpredictability of the rainfall (variability), and the scarcity of it (aridity), dominate.

8

The Sahel is a zone of grassland, scrub and thorny bush lying between the Sahara Desert and the wetter savannas to the south (Grove, 1978). The term has come to include, however, the semi-arid Sudanian savanna woodlands, which once formed a continuous zone to the south of the Sahel proper. The single short rainy season decreases in length from south to north. In the south, the rains can arrive as early as May and end as late as October; in the north they may not arrive until July and yet end in September. A lengthening dry season provides Koechlin (1997) with a basis for identifying four bioclimatic sub-zones from northern Sudan to sub-desert (Table 1.1).

Average rainfall, however, is a poor predictor of actual precipitation as it is subject to a coefficient of variation of 25–30 per cent. Older people can recall several calamitous failures of the rainfall during their own lifetimes.

The limitations of averages are well known. Variability occurs on four different dimensions, all critically important for farming (if not always for livestock) operations: (1) average rainfall varies over space, from south-west to north-east, as it does throughout the Sahel, predisposing ecosystems to certain crops and varieties; (2) actual rainfall varies in space over quite a short range, even between different plots farmed by the same family, as the distribution of showers is unpredictable (an intensive shower may, furthermore, contain quite a large proportion of the year's rainfall); (3) total annual rainfall varies from year to year (coefficients of variability reach over 30 per cent in the Sahel), dramatically changing the conditions for plant growth; and (4) rainfall is distributed very unevenly during a single season, so that a satisfactory total may nevertheless be poorly timed with regard to the growth cycle of plants, for example, when a sharp drought occurs during early growth, or in the grain-filling phase.

The designation, Sahel, tends to mask the diversity found within the region. In fact, it is spatially quite heterogeneous, and the complex interactions of society with environment demand a closer engagement than many studies have so far achieved achieved (Raynaut et al., 1997). At the micro-scale, diversity in the ecosystems is matched by diversity in agricultural practice and this in turn is linked to diversity among households (Piters, 1995). Diversity is

Table 1.1 Bioclimatic zonation of the Sahel

Sector	Annual Rainfall (mm)	Dry Season length	Vegetation
Sub-desert sector	200–250	10 months or more	
Sahelian sector	250–550	8–10 months	Dry steppe
Sub-Sahelian sector	550–750	7–8 months	Transitional steppe/ savanna
Northern Sudanian sector	750–1000	6–7 months	Sudan savanna

Source: Koechlin (1997)

also found in the economic strategies of households, migration, and the way these have an impact on women (David *et al.*, 1995). It is a major finding of our own study.

The area of north-east Nigeria which is the subject of this book is properly a part of the Sahel. It contains, in the Kano Close-Settled Zone, higher population densities than any of the CILSS member countries, but it faces a similar range of environmental problems. It is regarded, both by the Nigerian Government and by aid donors, as a region facing environmental degradation (having the European Community funded North East Arid Zone Development Programme), and has suffered a decline in annual rainfall, which, if averaged over the years 1961–1990, amounted to 8 mm per year (Hess *et al.*, 1995).

We hope, therefore, to reintegrate northern Nigerian experience into a debate which has been largely preoccupied with the francophone Sahel, on which there is a large and specialised literature covering many aspects of development policy and practice, founded in a substantial number of empirical studies. The relatively scanty Nigerian literature lacks the strong thematic focus of this corpus on Sahelian agricultural development, much of it sponsored in recent years by the Centre de Coopération Internationale en Recherche Agronomique pour le Développement at Montpellier (for example, Bosc *et al.*, 1990, 1992). Perhaps because of this, and the special conditions affecting northern Nigeria's political economy (the relative wealth of its government, high popula-

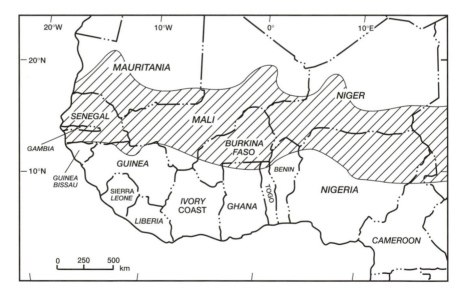

Figure 1.1 The Sahel.
Source: Redrawn from Koechlin (in Raynaut *et al.*, 1997: pp. 32–3). The northern limit shown for the Sahel is the boundary between semi-desert herbaceous or shrubby steppe and sub-desert vegetation. The southern boundary is the limit of Sudan *Isoberlinia doka* savanna and secondary forest-savanna mosaics.

tion densities and level of urbanisation), the responses of its smallholders to environmental and economic change may offer a foretaste of the future for other parts of the Sahel.

Population density and farming systems

In sub-Saharan Africa, population densities were, until recently, low by comparison with those of Europe and South or East Asia, except in some important but relatively small areas. The reasons for this state of affairs, in the continent where *Homo sapiens* originated and multiplied for millennia, and where evidence for culture is extremely old, are still imperfectly understood. They may have something to do with the special characteristics of African ecosystems, with the widespread occurrence of potentially negative factors such as aridity, variable rainfall, a scarcity of surface water, acid soils, and insect or parasite populations that are hostile to the health of humans or domestic animals. Whatever their explanation, the historical outcome of these low densities was that labour was (and in many places continues to be) the most limiting factor of agricultural production. Rural Africans struggled for millennia to evolve strategies for subsistence and environmental management that were effective with low labour inputs (Iliffe, 1995).

Economic history was centred on the reproduction, social control and productive management of labour, in order to harvest natural or husbanded wealth from the vast resources of land, water and natural vegetation. Many scholars have recognised that labour scarcity acted as an historical constraint on the creation of wealth in Africa. It was only when, in particular circumstances of history and geography, the productive capacity of labour could be concentrated under the control of a political hegemony, or into a small area, that this constraint could be broken. For example, the ancient kingdoms of Ghana, Timbuktu, Songhai and Kanem-Bornu achieved shortlived wealth through the political control of primary production and trade in extensive areas; at other times and places, dense rural populations grew up under the protection of centralised rule, for example, in the Hausa emirates, under the powerful chieftaincies of Yorubaland, or (paradoxically) under acephalous political organisation among the Igbo (Connah, 1988; Iliffe, 1995).

According to an indigenous view, growth in the population could only be a blessing, whether at the level of the household, the community or the state. More people meant more crop production, more time for tending livestock, and more opportunities for diversifying activities outside primary production. Non-agricultural communities, dependent on hunting and gathering from the natural ecosystem, had cause to control their own reproduction, unless territorial expansion was possible. Nomadic pastoralists, whose herds were found, in years of drought, to exceed the supply of fodder and browse, had to 'slough off' refugees to neighbouring regions whence agricultural expansion had excluded them. But for crop producing communities, a demographic strategy of high fertility was a necessary condition for escaping the bondage of underpopulation, as theorised by Boserup (1965, 1990).

In other continents, technological revolutions in agriculture substituted capital for labour on an increasing scale, after 1750 in Europe, and since 1945 in many south and east Asian countries. Productivity came to depend, to an increasing extent, on capital investment. The application of capital to agriculture released (or expelled) a majority of the population to feed urbanisation on a grand scale. Ploughing and carting with animal energy are, of course, much older than this; but with the major exceptions of Ethiopia and South Africa, sub-Saharan Africa was affected little by them until after 1945, and even today, change is slow and uneven (Pingali *et al.*, 1987; McIntire *et al.*, 1992). Paradoxically, the use of animal energy makes the fastest progress where population densities are high or rising. The introduction of fossil energy to replace labour, outside enclaves of commercial farming, is conspicuously slow.

In most countries of sub-Saharan Africa, therefore, agricultural capitalisation has lagged significantly behind the rest of the world. In all but a very few of them, more than half the population (and in some, more than three-quarters) live in rural areas and practise some forms of primary production (farming, livestock husbandry, fishing, hunting or collecting). In these productive systems, even where some mechanisation has occurred, labour continues to be the key to productivity, and most of this labour continues to be *family* labour. The social institutions whereby this labour is controlled and managed continue to determine the distribution of benefits from agriculture, even in rapidly evolving systems where market crops are grown (Berry, 1989).

Changes

There have been some fundamental changes in the relationships between population, labour and natural resources. First, improvements in public health, coupled with some advances in incomes, welfare and diet, have brought about a fall in mortality rates and an increase in the rates of natural population growth during the second half of the twentieth century. However, rather than reduce their fertility, households in sub-Saharan Africa have embraced the larger family with enthusiasm, seeing in it their opportunity to increase the number of workers and diversify the sources of income for the household. Only in the last few years have falling levels of fertility been highlighted in the literature (Blacker, 1993; Gould and Brown, 1996).

Second, increasing numbers of claimants to land and other natural resources have raised labour to land ratios to levels (in places) which were previously unknown, and the demands they make on soil fertility, biomass, faunal populations and surface or groundwater are believed by many to compromise the renewability of these resources, at least in the medium term (Turner *et al.*, 1993).

Third, some pre-colonial institutions of social control have been eroded, as the autonomy of the individual *vis-à-vis* the household, and of the household *vis-à-vis* the community has been strengthened, having negative implications for the use of collective labour and the management of common property resources.

Fourth, urbanisation and the growth of markets, which have both gathered increasing momentum during the last two decades, together with improvements in transport infrastructures and in the real costs of travel, have widened the economic opportunities available and the geographical horizons that can be exploited by rural individuals (Snrech *et al.*, 1994).

Given such far-reaching changes, what is the position, as we approach the centenary of colonial intervention in much of Africa, of the rural household? How does it manage its endowment of resources – natural, economic, techno-logical and social? In particular, what is the position of labour in the strategic and tactical (short-term) choices made by households as they confront the many challenges facing them? How far have increased numbers of people per hectare provided an opportunity for productive transformations of natural landscapes or for enhanced livelihoods for rural households, and to what extent have they induced degradational trends or impoverishment? Our transect allows us an opportunity to look at these themes at a micro-scale, and longitudinal analysis will permit them to be understood in a time perspective.

The thesis of this book is that, on the basis of the evidence analysed here, the 'Sahelian crisis' of degradation can be contained, and that in doing so, the resources of rural communities themselves will play a much larger part than is usually assumed. Pre-eminent among these resources is the labour provided by a growing population which, in drawing on a wealth of indigenous technical experience and the best of introduced practices, can create, through an incre-mental and 'indigenous' intensification of agriculture, more sustainable produc-tion systems. Gradualist rather than transformational expectations should therefore underpin the policies of governments and donors, policies which need to be founded both on an improved understanding of the diversity and the dynamics of primary production systems, and on a recognition of the need for unimpeded economic integration between the Sahel and West Africa as a whole.

2

DIVERSITY, FLEXIBILITY
AND ADAPTABILITY

In this chapter, we shall develop a simple model which expresses the relationship between the key constraints of rainfall, bioproductivity, labour and capital, households' responses to these constraints through managing diversity, flexibility and adaptability, and the outcomes of such management. We will outline our method of analysis of the linkages between the management of labour and natural resources, linkages which are the focus of this book. We revert to the biogeographical definition, which includes a substantial portion of northern Nigeria and smaller parts of Ghana and Cameroon.

Degradation, or desertification as it is commonly called, is another apocalyptic feature closely associated with images of the Sahel. Indeed, the Sahel drought was instrumental in bringing about the United Nations Conference on Desertification in 1977, and has figured prominently in action plans and debates on that subject ever since (Adams, 1990; Swift, 1996). Large sums have been invested by donors in anti-desertification projects which are generally agreed to have had little impact.

Constraints and responses

Four key constraints

Enough has been written above to show the importance of *rainfall* – its scarcity (in most years) and its variability – for Sahelian farmers and livestock producers.[1] In the following chapters, we shall explore the implications of these characteristics of the rainfall for the farming systems. We consider rainfall to be an exogenous variable; that is, one which lies entirely outside the farmers' control. Closely linked to the rainfall is the *bioproductivity* constraint (by which term we refer to plant bioproductivity and the condition of the soil which sustains it). This constraint is not exogenous, as the farmer or livestock producer can influence it by adopting either sustainable or unsustainable management practices. The extent of his or her ability to do so is, however, hotly debated. *Labour*, as we have emphasised already, is a third major constraint, and one over which households can exercise influence through their fertility behaviour,

14

through social institutions for sharing labour or appropriating the labour of others, or through hiring it. However, as we enlarge the scale of analysis from the household to the community, the village or the region, labour sharing between households and exchange tend to cancel each other out (unless there is migrant labour), and the influence of fertility on the natural increase of the population becomes the uppermost determinant of labour supply. Finally, there is a *capital* constraint. The poverty of Sahelian rural households (as a whole) is not disputed, and this poverty restricts the possibilities for technical change in farming and livestock production.[2]

Responses

In most African drylands, including the Sahel, labour is the only major constraint that has been consistently relaxed over time (owing to population growth). This fact has assigned it a dominant role in forcing change. Initially, additional labour enables a production system to overcome the disadvantages of 'underpopulation', especially a shortage of labour; later, continued population growth provokes a scarcity of land or other natural resources. Tiffen *et al.* (1994) identify three types of response to population pressure: (1) intensification of resource use *in situ*; (2) diversification out of primary production into other income-earning activities; and (3) migration to other areas where natural resources are still available.

Social communities construct their livelihood systems in response to these four major constraints and the opportunities available to them, which group themselves into these three main forms of response. However, this simple characterisation fails to pursue several questions about how and why particular decisions are made. We propose to discuss these questions in terms of three variable properties of livelihood systems (and more particularly, natural resource management systems), each of which it is in the interests of Sahelian rural communities to maximise:

1   Diversity in resource endowments and available options;
2   Flexibility in the day-to-day management of these resources and options;
3   Adaptability over the longer term in defining new or altered systems.

This categorisation represents a development of ideas initiated elsewhere (Adams and Mortimore, 1997). Six dimensions of flexibility which we recognised are: (1) in the use of farm labour; (2) in the use of cultivar biodiversity; (3) in the use of economic plants; (4) in field location; (5) in the use of grazing resources; and (6) in the choice of livelihood strategies. Here, we distinguish between flexibility and adaptability. We use the term *flexibility* to refer to short-term decisions about the use of resources, and therefore include decisions about the deployment of family labour on a weekly or even daily basis. We use *adaptability* to refer to longer term decisions and to inter-year adjustments and responses.

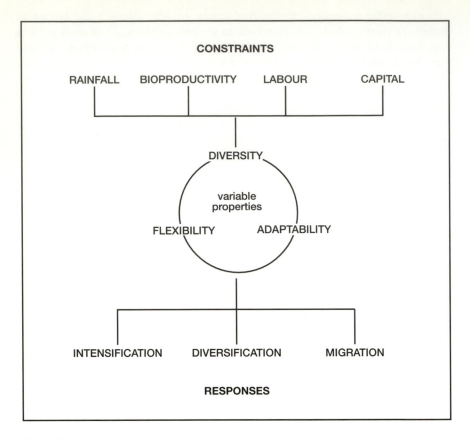

Figure 2.1 Constraints and responses.

We also introduce the concept of *diversity* in resources and in the options available to households, which defines the scope of the responses that can be made. The rationale of diversification is fundamental and relates to the spreading of risk, which is the key to the economic survival of the household and its members. In the Sahel, risk is dominated primarily by rainfall variability and its ecological and economic consequences. Risk management is achieved through maximising opportunities for the accumulation of wealth of either individuals or households, which often means tapping wealth flows in other regions or sectors through engagement in off-farm and (often) out-of-village economies.

The relations between these three response components, and the constraints identified earlier, are shown in Figure 2.1.

The goals of dryland households need to include maximising the variables diversity, flexibility and adaptability in their production and livelihood systems. We briefly discuss these attributes below. In doing so, we hope to expose the complexity of a smallholder livelihood system which combines the management of natural resources and economic livelihoods with the pursuit of reproductive and social goals, in a context of risk and environmental change. It is this 'wholeness' or integrity which defines smallholder systems throughout the world, according to Netting (1993).

Diversity of resources and options

The resource endowment of a household, at a point in time, consists of four dimensions.

Natural resources

The natural resource endowment of a household is the productive capacity of soil and other living resources to which members of the household enjoy rights of access, both to work and to enjoy the benefits. Such endowments are variable along three dimensions: socially (some households or individuals within households have different endowments from others); in time (both the natural productivity of such resources, and people's rights of access to them, change year by year); and in space (there are sharp differences between places, between communities and between micro-niches, even on the same farm). At a point in time, endowments are 'given'.

Diversity is exploited in order to spread risks, and to enlarge opportunities for economic gain. Contrary to a common misrepresentation of the Sahel as homogeneous, smallholders actively 'cultivate diversity'. For example, soil and soil moisture variability in space is exploited by diversifying cropping patterns within the farm or field; differences in natural vegetation communities are efficiently grazed by diversifying the species in domestic herds; an assortment of multipurpose trees is left standing on farms whose economic value can be realised (for food, fodder, medicines, construction materials) at different times of the year.

Economic resources

The economic resource endowment of a household includes the labour and capital available to it either from the family or through social or commercial means of appropriating it from outside; it also includes access to markets (for selling output, buying inputs and for earning off-farm income). It is expressed in the options available for strategic choice. Like natural resource options, economic options are distributed unequally – socially, temporally and spatially.

Diversity, rather than specialisation, is the goal in managing households' economic resources. For example, farmers mix their crop enterprises and cultivar lines, keep livestock along with farming, and diversify animal herds; household members (including women) try to have at least one off-farm source of income; and most households send one or more members away from time to time to earn income elsewhere. It is a recognised principle that the more diversity, the better.

Technical resources

The technical resource endowment of a household includes a range of options available to it in farming, livestock husbandry and off-farm activities, whether carried on at home or away. There are levels of 'availability' or access to technologies which have practical importance for households. A first group includes those for which both the necessary knowledge or skills, and the tools or equipment, are already present. They form a part of the household's normal operations. A second group includes those to which access can be gained relatively easily by exchanges within the community. This often characterises off-farm or migrant trading options. A third group is known but lies beyond reach on account of the capital resources needed to gain access. Thus the concept of 'options' should be used with care when applied to particular households.

A wide range of technologies is applied to the natural resources of the Sahel. These include the following: soil management technologies (such as conservation structures, drainage, various ways of returning organic matter to the soil (manure, compost, burning residues), fertilisers, various ways of preparing the soil (ploughing, hoeing)); weed management technologies (hoes, ploughs, herbicides); pest management (restricted to a few local preventive practices and insecticides); harvest management (methods of cutting, drying, transporting and storing crops); animal management (methods of breeding, feeding and health care); fodder management (uses of residues, natural grazing, browse of a variety of trees); water management (wells, boreholes, pools and systems of watering herds and rationing when necessary); irrigation management (pumps, manual lifting devices, channel and flow control, flood recession control with bunds, etc.). Additional technical knowledge is used to gain access to economic resources, for example, transport (donkeys, horses, cycles, motor vehicles); skills to operate equipment used in off-farm activities (mat making, sewing, butchering and a host of others), often passed down from father to son or from mother to daughter; knowledge of markets; networks of information concerning employment or trading opportunities elsewhere. The list is almost endless, and the conclusion obvious: that a 'normal' household optimises a range of technical resources, access to which varies between households.

Social resources

The social resource endowment of a household includes the age, gender, health and skills of household members, their availability for economically productive work under ruling social *mores*, and the place of the household in the social community, which in turn determines its access to sharing institutions and

18

allocative procedures. No household can exist in isolation from the community, which at times makes demands on its members (for example, in cooperative work groups, or in time taken from farming in order to observe socially approved rituals) and at others secures its rights (for example, to the use of disputed land). Production is, indeed, embedded in the social fabric where a continuing negotiation takes place between the individual and the community. The fluctuating course of such negotiation, which is pulled in opposite directions by custom and market forces, defines the way in which the community as a whole responds to changing economic and environmental circumstances.

The diversity of inventories of natural resource, technical and economic options *actually available to the household at any point in time* may vary considerably, for example, with household economy and composition, and with prevailing environmental conditions. It varies between households within one place (as a function of wealth, position in the household reproduction cycle, and contingent events such as illness, misfortune or bereavement), and also varies over time (as household compositions evolve and their economic fortunes change, and as external economic and environmental factors have influence – such as in successive years of low rainfall). However, while the diversity of resource endowments can and will change for each household and between households, it is at any given point in time a static measure of potential for making decisions: a description of the state of the resources and opportunities available to the household.

A fundamental question is: How much choice and how much compulsion do dryland environments offer or impose?

Flexibility

Whereas 'resource endowment diversity' is a measure of natural, economic, technical and social resources, including the capacities to use them, *flexibility* describes active decisions taken within households about resources. Flexibility represents the capacity of those decision makers within a household to exercise short-term choice, selecting agricultural or livelihood strategies, 'coping' with economic or environmental circumstances and making the most of the diversity of opportunity available to them. Flexibility is demanded, in particular, by the need to deal with unpredictable events like a dry period during the rainy season (which may require replanting of crops, or decisions about alternative income or food), events which, turning up unannounced, call for rapid responses. These may follow familiar or unfamiliar patterns. Flexibility is also needed in order to respond to unpredictable economic events, personal or health setbacks, and other environmental variables such as pests.

Flexibility is more than mere choice, though obviously the range of choice determines its limits. It is more than economic opportunism, as social and psychological variables are involved. To some extent it is built into the system of production; thus the six broad types of flexibility identified above may be said

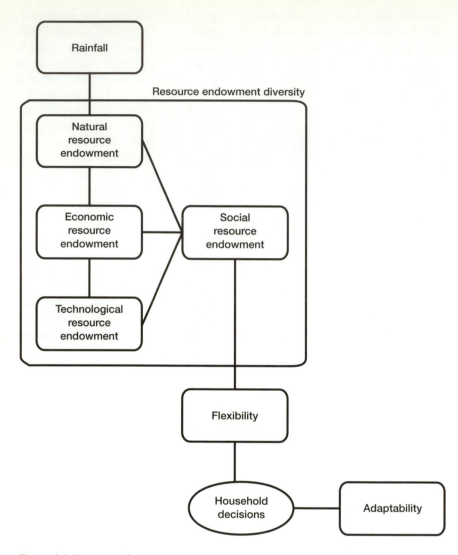

Figure 2.2 Diversity of resource endowment.

to characterise the livelihood systems of our four communities in north-east Nigeria. But it is well known – and recognised by rural people – that flexibility in the system does not necessarily imply equal flexibility in all households, where (for example) old age, or a shortage of family labour, or an absence of famil- iarity with technical options, or chronic sickness, or an inability to estimate the opportunity costs of labour in alternative occupations, or a host of other factors at the individual level may severely constrain the possibilities. But it is clear that Sahelians have to stay on their toes. They represent the end of a spectrum whose opposite end is the predictable, equilibrial world of a plantation worker in the rain forest.

In Figure 2.2, therefore, flexibility is shown in a mediating position between resource endowment diversity and the decision-making process. Many examples can be given. In farming, decisons about what to plant, and where, are reviewed in the light of the previous year's performance, of expected rainfall, of insect pests, of price movements, of the amount of food still remaining in the family granary, and of the labour available. In livestock production, decisions (whether to buy or sell, where and when to graze) are affected by the previous year's rainfall, the condition of the natural vegetation, the amount of residues available, prices (of fodder and of animals), or other household priorities. Decisions about livelihoods are affected by the household's food sufficiency, the age and fitness of the male population, experience of employment or trading opportunities elsewhere, and so on.

Adaptability

Farmers and other resource users in the Sahel do not simply respond to exogenous and endogenous change reactively within each growing season. Their decisions, when considered in sequence, are cumulative and purposive, and therefore have a longer term significance which reflects their understanding of longer term and larger scale changes. Denial of such a perspective has led many writers to assume that 'planning horizons', especially of the poorest, are short, and that 'time discount rates' on investments are therefore high. This is not to claim that all decisions about managing natural resources have a positive effect on sustainability. But they are both rational and 'pathway-dependent'; that is, each decision in a chain is influenced both by goals and by previous decisions.

This sequential element, we suggest, can best be captured by the term *adaptability*. Adaptation best describes the longer term restructuring whereby the Sahel has sought to overcome the impact of the environmental and economic setbacks it has experienced since the 1970s and over the longer term (Mortimore, 1989). Adaptability in the household's system of managing its resources from year to year can transform its production systems (or its mix of livelihood-generating activities) in response to changing conditions. It is, in a sense, a valuation of all the household's decisions taken within the year, which can be expressed on a scale from low to high (Figure 2.3). Although qualitative rather than quantitative in nature, this scale measures 'performance' in the Sahelian setting. It determines the starting point for the following year's flexible decisions. Over a series of years, a highly adaptive household must become stronger and a maladaptive one weaker.

Adaptation is quite likely to be episodic or bunched in time, when circumstances demand certain kinds of action (such as in drought). It may be clustered in space, as the decisions taken within households differ from one place to

21

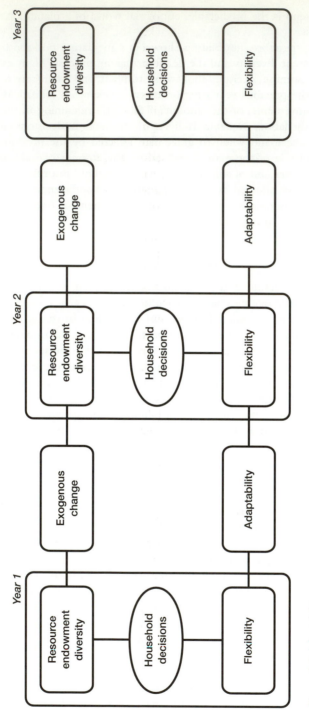

Figure 2.3 Adaptability.

another (such as between villages) because of micro-environmental patterns. It may be clustered socially, as the decisions taken within different households in one village differ because of their social or economic situations. As these variables operate year after year, significant differentiation builds up.

Adaptability is exemplified in the following widely attested changes in farming. There has been an evolution of reservoirs of indigenous technical knowledge, plus exogenous technical knowledge (ITK + ETK) at different places and under differing conditions. Over time, through selection, new technical options (for example, cultivars) and new modes of management (for example, of trees) have emerged. Communities have capitalised their villages (for example, with permanent structures), fields (for example, with fencing or hedging), farming (for example, with ploughs), and water supplies (for example, with new wells). Modes of exchange and transfer of rights of access to natural resources have changed in relative importance (for example, increases in sales of cultivated land) in order to accommodate a need to increase the efficiency of land use. Institutional change has accompanied such developments (such as the decline of cooperative labour with increases in labour hiring).

A rationale for adaptability in household livelihood strategies is found in the need for sustainability of livelihoods and natural resources, and its evidence is found in systems in transition under pressure of change. There is a growing literature on 'indigenous knowledge', and on the capacity of farmers and other resource users to respond to opportunities and threats (e.g. Richards, 1985, 1986; Reij *et al.*, 1996). There never was such a thing as 'traditional' management of natural resources. Practices and goals are continuously revised in the light of *current* conditions. Conditions are, however, changing much faster today than they have in the past. Adaptability has always been a feature of successful livelihood strategies in the Sahel, but in the late twentieth century it has acquired a heightened importance and has involved a particularly wide range of activities as livelihoods have been influenced by environmental, political, economic and social change.

Research perspectives

The literature on Sahelian rural systems is a rich one – in relation to their populations, more so in some francophone countries (such as Senegal and Burkina Faso) than in northern Nigeria (see, for example, Bosc *et al.*, 1990). Why add to it? Notwithstanding its apparent abundance, it can still be said that in relation to the known diversity of the Sahel, the number of systems recently described or analysed is still woefully inadequate as a basis for generalisation. Drawing selectively on only a small fraction of this literature, we can identify five themes, important for our own objectives, before showing how this study attempts to innovate.

Several major studies of village or community systems in the Sahel, broadly defined, have made use of an ethnographical–geographical tradition which incorporates a holistic frame of reference, in which society, resources and livelihoods are linked, and which attempts to capture indigenous perspectives on matters which began early to attract technical judgements from outsiders (such as shifting cultivation or livestock breeding). We can exemplify such studies by citing those of Mainet and Nicolas (1964), of Nicolas (1975), and of Raynaut (1973) in the Maradi region of Niger, of the Serer of Senegal by Gastellu (1981), of Gallais (1984) on the Niger Inland Delta, of Hammond (1966) and Marchal (1983) in Yatenga, Burkina Faso. Not only farming but also pastoral systems were the subjects of such studies, such as those of Stenning (1959) in Nigeria, Dupire (1962) and Bernus (1981) in Niger, and Barral (1977) in Burkina Faso, and mixed systems, such as those by Hopen (1958) in Nigeria and Delgado (1979) in Burkina Faso.

The best ethnographic studies were the product of 'participant observation' during prolonged periods of residence in the community or region. We have chosen to stick with this method, in preference to the currently popular rapid 'participant rural appraisal', both because our objectives called for data series covering several years, and because we needed to aim at a sympathetic understanding of the interconnections linking natural resources, livelihoods and family behaviour. The fieldwork for this study was carried out by Nigerian researchers, whose own language and culture put as little distance as possible between them and their host communities – with whom, indeed, many friendships were formed. We also find common ground with the holistic frame of reference of the ethnographic studies; Hammond, for example, analysed an 'integration between the natural environment of Yatenga and the technological adjustment which the Mossi have achieved to it' (1966: 205).

A second major theme is a tradition, more commonly found in the French literature than the English, of studying natural resources at the village level through the cartographic analysis of the *terroir*, given expression in pioneering studies by Pélissier (1953) and Sautter (1962). The *Atlas des structures agraires au sud du Sahara* included several Sahelian studies, including Lericollais (1972) on Sob, Senegal; Savonnet (1970) on Pina, Burkina Faso (1970), and Hallaire in north Cameroon (1971). The subject of land tenure has recently come to occupy a larger place in theorising and debate on economic development, drawing heavily on such studies (see le Roy *et al.*, 1996; Lavigne, in press). We do not explore issues of land tenure in this study. Nevertheless, the conception of natural resource management which we have used begins with the territory of the rural community, its spatial organisation and the social distribution of access to it. We too have constructed working cadasters of arable land in the field.

A third research theme is the more focused study of agricultural economics at the village or local community level which, in northern Nigeria, we can trace back to the pioneering investigations of Forde (1946) on the Hausa, Smith (1955) on Hausa communities near Zaria, and Luning (1963) on a Hausa locality in Katsina. In neighbouring Niger, Nicolas (1965) turned from ethnography towards economic organisation among the Hausa of Maradi. In the Gambia, Haswell (1953) carried out one of the earliest studies in the village of Geneiri. Driven by demands from economic planners for strategies of programme intervention in what were seen as 'traditional' and conservative farming systems, the theoretical and methodological rigour of agricultural economic studies, as well as the resources available for them, escalated. Input–output studies based on much larger data sets, and hypothesis testing using more sophisticated tools such as production functions or marginal analysis, seemed to become the norm. At the level of a single village, Toulmin (1992) showed how the economics of farming and livestock production can be analysed in a robust and inclusive theoretical framework.

None of these initiatives in agricultural economics were more ambitious than that begun by the Rural Economic Research Unit of Ahmadu Bello University (RERU) at Zaria in Nigeria, a project which became a kind of baseline for subsequent farm studies in northern Nigeria. Later to be re-presented as a progenitor of farming systems research (FSR) in Nigeria (Norman *et al.*, 1982), this elaborate programme of input–output studies involved no less than twelve villages in four sets of three, arranged at differing population densities and distances from town. The four sets were near Sokoto (Goddard *et al.*, 1971; Norman *et al.*, 1976), Zaria (Norman, 1967, 1972), Bauchi (uncompleted), and Ilorin – distributed among the Sudanian and Guinean agro-ecological zones. The project, not surprisingly, had to depart from the plan in some respects, and was slow to reach publication (for which it was criticised by Hill, 1986). Nevertheless it produced an unprecedented volume of data on large numbers of smallholdings in the late 1960s. As a result of this work, the economic rationality of farmers' behaviour, and its susceptibility to standard analytical methods, were established beyond any lingering doubt. Methods of quantifying household and farm level data, of sampling in the field and of quality control over periods of data collection prolonged over many months, were well tested under northern Nigerian conditions. Other workers, ourselves included, have benefited.

The RERU studies took a step forward from the 'single village' study to adopt a comparative method, in order to test hypotheses, notably on the relations between intensification and population density, and between economic activity and distances to markets. Though this thrust was later abandoned in the few publications that resulted, theoretical investigations based on comparisons of production systems, using standard protocols to control variability, have been tried on a continental basis in recent years – on farm mechanisation (Pingali *et al.*, 1987), on the integration of crop and livestock farming (McIntire *et al.*, 1992), and on the intensification of agriculture (Turner *et al.*, 1993). This essentially geographical strategy in theoretical investigation provides us with our fourth theme, and it has been built into our research design, using demographic, ecological and market variables (explained in the next section).

Agricultural economic studies depend on data sets which are expensive to collect and are nearly always restricted to a single year or farming season. They have been less successful in analysing economic differentiation, as it is a time-related process; indeed the input–output studies in northern Nigeria told us less of the nature of poverty than Hill in her studies in Katsina and Kano (Hill, 1972, 1977), though much smaller in scale. Neither could they confront the environmental and economic changes which were taking place. If classic studies of village systems had described an 'ethnographic present', many agricultural economic analyses can be accused of uncritically portraying an 'economic present' or, even more amazing (in the light of current obsessions with global change), a stable environment.

After the Sahel Drought of the 1970s, the days of this last assumption were certainly numbered. In its aftermath, another substantial corpus of field studies was carried out by the University of Bordeaux, this time in the Department of Maradi in Niger. Far from accepting an unwritten assumption of equilibrium, as some earlier studies in Nigeria had done, this work took as its starting point the social and environmental stresses that had been identified during the years of drought. A technical study of natural resources and their management (Koechlin, 1980) was linked with social and economic analyses of production systems, migration and livelihoods (Grégoire and Raynaut, 1980; Raynaut, 1980), and the regional synthesis was based firmly on detailed studies at the village level (e.g. Raynaut, 1977a; Grégoire, 1980). Where uncontrolled expansion of farming and woodland clearance were proceeding apace, driven in part by population growth, another drought cycle in the early 1980s appeared to provide ample justification for a gloomy prognosis (Raynaut, 1997b). With these scholars we accept (as our fifth theme) the absolute necessity of designing research to take account of the contemporary context of rapid and controversial environmental change.

Change has now risen to the top of the global as well as the Sahelian agenda. In view of what has already been written, it is not surprising to find increasing recognition that, for understanding economic and environmental change, studies conducted on a single-year basis are inconclusive. It can also be convincingly maintained that diagnostic surveys of degradation, which relied traditionally on technical interpretations of soil or vegetational indicators, may also be an unreliable guide to future trends, not only on account of the inter-annual variability which characterises the dynamic of the Sahelian environment, but also because of the adaptive potential of its management by millions of smallholders, whose impact on plants, animals, soils and water is all-pervasive. As the shadows of the pioneering village studies lengthen, and individual researchers grow older, the possibility of using follow-up studies to investigate change becomes worthwhile. Haswell (1963) provided an early trial of this kind of approach in the Gambia, and later went on to extend her time horizon from a decade to a quarter-century (Haswell, 1975). A study of change in the agrarian systems of the Serer has been published (Lericollais, 1979), and in that of the Kano Close-Settled Zone (Mortimore, 1993a). Elsewhere in dryland Africa, Tiffen *et al.* (1994) reversed conventional interpretations of change in

Machakos District, Kenya, with a multi-disciplinary analysis of some data series that spanned 60 years. Such a 'surprise' in research outcome suggests that it is important to get the methodology right. Air photographs, which began to become available around 1950, afford an opportunity to measure changes in some vitally relevant variables (Turner, 1997).

Finally, we return to the question of variability introduced earlier. Studies in the drylands of Africa have stressed the need to recognise variability and dis-equilibrium as fundamental properties not only of rainfall but also of terrestrial ecosystems, and opportunistic strategies as a necessary cornerstone of management for small farmers and stockraisers (Sandford, 1983; Behnke *et al.*, 1993; Mortimore, 1989, 1998; Scoones, 1994). We can, with comparative ease, tell planners that it is unrealistic to assume constant, average or equilibrial conditions. Abundant evidence comes to hand and cheap points can be scored by drawing attention to the many project failures. Risk has long been a subject of analysis in agricultural economics, explored by Toulmin (1992) in a Bambara community in Mali, and in a broader disciplinary context by Scoones *et al.* (1996) in Zimbabwe, and by Davies (1996) in the Sahel. An economic framework of risk analysis can only investigate risk in terms of incidence, choices of avoidance strategies, outcomes and costs. We would like to go further.

Of the five themes briefly introduced in the foregoing discussion, it may be said, first, that the value of a holistic framework is regaining recognition after several decades of increasingly specialised disciplinary research; second, that the *terroir*, which in recent years has formed a basis of programmes of political decentralisation in several francophone countries, helpfully redirects attention from output and income maximisation to the sustainable management of natural resources; third, that agricultural economics, while demonstrating in more and more detail that what smallholders do is rational, offers (alone) a framework too restrictive for understanding the multi-sectoral and time-dependent dimensions of household livelihoods; fourth, that research designs using a comparative method offer opportunities to test certain theories in a continent where geographical diversity is of the essence; and fifth, that long-term economic and environmental change not only help us to understand what is happening at the local level, but to link localities and communities with global change.

Within this framework, we aim to analyse systematically, and in sequence, the *adaptive behaviour* of households: the allocation of scarce resources (labour in particular) among contending demands. We aim to quantify these decisions. Policies for the drylands need to take account of how smallholders manage their constraints and opportunities, and find ways to support them and enhance good management.

Research design: the villages

The research on which this book is based involved parallel investigations in four villages in north-east Nigeria. These lie on an environmental and socio-economic

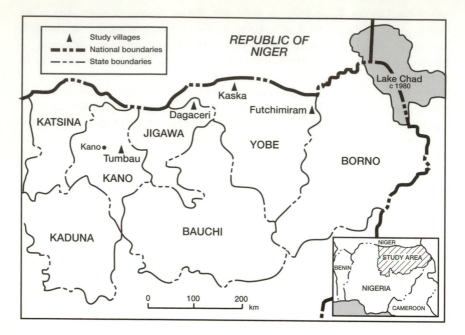

Figure 2.4 North-eastern Nigeria, showing the four study villages.

gradient of decreasing rainfall, population density and farming intensity; they also lie at increasing distances from the regional market at Kano.

In the Sahel, bands of diminishing average rainfall succeed one another northwards, and to take this characteristic into account, four sites were selected: one Sudanian (Tumbau, Kano State), one transitional (Dagaceri, Jigawa State), and two Sahelian (Futchimiram and Kaska, Yobe State). The four sites are situated on this gradient of declining average rainfall from a thirty year mean (1961–1990) of over 600 mm to leass than 350 mm. Their average rainfall, during the growing period (between May–June and September–October) in the five years from 1992 to 1996 was 533 mm, 360 mm, 326 mm and 301 mm respectively. The locations of these villages are shown in Figure 2.4.

A gradient of population density follows that of the rainfall, in direction though not quantitatively correlated with it. Tumbau lies in the fertile Kano Close-Settled Zone, which has higher densities than anywhere else in the Sahel ($400/km^2$). Dagaceri lies in an area of former sand dunes settled mainly within the last 100 years ($50–70/km^2$), Futchimiram is located in the extensive wood-lands of former northern Borno ($20–30/km^2$) and Kaska lies among still active dunes and open grasslands on the border of Niger ($<15/km^2$).

Reflecting these variables, the overall intensity of rainfed farming declines from Tumbau to Kaska, though the complexity of land use is such that no simple correlation should be inferred. This matter will be taken up in Chapter 6.

28

On the face of it, another gradient (of market access) is reflected in the design, and can be expressed in increasing distances from the regional market of metropolitan Kano. But such is the ubiquity of the urban and rural periodic market system in Nigeria that mere distance has less significance than some other variables (for example, transport). It is the nature of market participation, rather than its level, which changes according to geographical location.

Labour, flexibility and adaptability: a method

Sustainability is a concept much discussed in development, but hard to define (Lele, 1991). Sustainability in a production system involves three dimensions: maintaining or enhancing its biological productivity (that is, 'ecological sustainability'); maintaining or enhancing its economic productivity – the benefits flowing from the system and from other linked economic activities ('economic sustainability'); and for equity in the enjoyment of these benefits, both among the present generation and between present and future generations ('social sustainability': Munasinghe, 1993). The challenges to sustainable agriculture, in what Pretty (1995: 19) refers to as the 'diverse and complex lands' (in which we would give a central place to the Sahel) are considerable, but by no means insuperable, as indeed an open-minded look at the past performance of African agriculture shows (Wiggins, 1995).

Given the uncertainty of their environment, the livelihoods of Sahelian farmers and livestock keepers depend on successful management of their resource endowment, and hence a very specific concern for environmental sustainability. At the same time they have to achieve economic survival, and depend on (and maintain) the social relations of production and reproduction. Their livelihoods are therefore inextricably dependent on sustainability in a quite fundamental way. The chief strategy for achieving environmental sustainability is through the exploitation (and maintenance) of the diversity of their resource endowment. Flexibility in making decisions, and adaptation of their systems in the longer term, are their means of achieving this goal.

The deployment of their household labour (which includes family and shared or hired labour from outside) is critical. Choices between different economic activities (for example, farming, livestock keeping, trading or crafts) are constrained by their availability, both in the village and to the individual household (and hence by the natural, economic or technical resource endowment); however, they are essentially choices about the things that people do in the course of the working day (and night). Similarly, choices of which crops to plant, how to manage them (the timing and intensity of weeding or the amount of fertiliser), and of which fields to lavish most time on, are all decisions about people's work. Decisions about resources therefore involve, in a very fundamental way, decisions about labour allocation. In this book, the issue of decision making about labour is used as a way to explore livelihood and subsistence strategies. The question of how and where people work provides an insight into the plight, and the power, of poor or richer households. More than that, however, an analysis of labour management at the household level is expected to provide an insight into the critical process of labour intensification, whereby growing populations come to terms with the need to sustain the productivity of their natural resources.[3]

Our approach, therefore, is to analyse the day-to-day management of labour by a range of households in four villages (whose natural and social conditions vary) in the Nigerian Sahel, over four consecutive years. In doing so, we have tried to encompass a broad range of ecological and demographic conditions and to consider how the systems respond to the variable rainfall over four successive growing seasons. By controlling for all these variables we hope to understand the dynamics of labour use. More specifically, the use of quantitative time budgets allows us to explore decisions about the allocation of household labour: in a sense, to start to explore one vital part of the anatomy of decision making that is the core of sustainable livelihoods.

An outline now follows of the methods used in the study.

Characterising systems and inventorying options

A fundamental task in commencing field research in a little known area is the characterisation of the farming-natural resource system. While there have been useful general accounts of agriculture in Hausaland, and studies in other parts of northern Nigeria (e.g. Hill, 1972; Norman et al., 1982; Watts, 1983), it is vital to recognise the local diversity of farming systems. Both within Nigeria and elsewhere, there is a tendency to assume that a small number of case studies can be used repeatedly to make generalisations over large areas. The farming systems of the four villages referred to in this study are described in detail in Chiroma (1996), Ibrahim (1996), Mohammed (1996) and Yusuf (1996).

Choice is focal in our model, and so inventories of technical and economic options were needed at the outset. These were compiled from interviews with selected informants in all four locations, and included mapping and classifying soils, inventorying natural biodiversity and plant and animal domesticates, investigating livelihood options and evaluating the technical and management options available in each community.

A longitudinal framework

Adaptive response to environmental change must obviously be analysed in a longitudinal framework, and failure to do this has been responsible for many superficial or misleading diagnoses of degradation. Two time dimensions are incorporated into the present study. First, observations were sustained during four consecutive growing periods of 1993 to 1996, in order to take account of inter-year variability. Second, greater depth was obtained by reference to archival studies of environmental management in the first 20 years of the present century, and through sequential air photo interpretation of land use change since 1950 (the year when the first remotely sensed data were recorded in northern Nigeria: Turner, 1997). The longitudinal data supply the basis for the model outlined earlier in this chapter.

Collaborating households

The main focus of this study is on the ways in which people allocate their labour power. Labour is the major controllable constraint affecting output in a system of hand technology, where so much falls beyond the farmers' (or stockbreeders') power to change. This is as true in intensive farming systems as in more extensive ones.

The household was chosen as the largest unit in which individual decisions can be understood. At the village, district or higher levels of scale resolution, individual decisions are necessarily aggregated, and causal relationships have to some extent to be inferred. At the individual level, on the other hand, collective action tends to be ignored. Households, notwithstanding much evidence for their internal complexity, not to mention the difficulties of defining them, still do operate as coherent management units in many areas of direct relevance to natural resource management – the family farm, the family grain store, the household livestock herd, to mention only the most important (Netting, 1993). Pursuit of this question at the household level demanded high frequency monitoring of the households during each agricultural season. The acquisition of these data was assigned a high priority in the execution of this study.

Depth was chosen in preference to breadth, which entailed a small sample. The reasons for this were twofold. First, a high level of collaboration was necessary in data collecting which was both intrusive and long drawn out – without goodwill, even personal friendships with the researchers, continuity could not have been maintained. The requirements of our model could not have been met had the collaborating households changed from year to year. Second, the nature of the information required did not permit a survey approach, as relationships were to be analysed on the basis of individual decisions and not inferred statistically. Of the 47 households initially recruited into the study, only five dropped out (all in Futchimiram, where there was a break in data gathering in 1994), later to be replaced by three more. The households were selected on the basis of a preliminary ranking according to wealth indicators, qualified by willingness to collaborate in the study.

Questions of method: time budgets

The only practical way of measuring labour input is by time. In our fieldwork, researchers collected data on the time given by each member of our collaborating households to different tasks. They visited each household at intervals of two or three days during the growing period, with longer gaps (when their

31

absence from the village was unavoidable) reconstructed by recall. For every person in the household, their place of work (or other activity) and task were recorded for the morning and afternoon of each day. The intention was to cover all activities, including rest and sickness, irrespective of location, in order to obtain an understanding of trade-offs and perceived opportunity costs of alternate uses of labour time. Women's and children's activities were included in order to quantify their contribution to farm work and other activities.

Five questions of method surround the use of labour time as a measure of intensity.

1 How much is a day's work worth?

First among these is the variable *value* of an hour's or a day's work by different persons unequally endowed with physical strength, skill or commitment. We attempt to deal with this problem by attaching weights on the basis of age and sex in order to reach a better evaluation of the relative importance of individuals' work contributions.

These weights were as follows:

Adult male (15 years to elderly)	1.0 unit for a full day
Adult female (15 years or marriage to elderly)	0.7
Boy, girl (8 to 14 years or marriage), elderly person	0.5
Child (4 to 7 years)	0.3

However, the weighting given is arbitrary, as the values have not been put to empirical test. Variability among individuals of the same age or sex, and between days or times of the year for the same individual, is not taken into account. Yet the essence of smallholder farming is that such differentials have an impact on productivity. Smallholders are not merely machines of low calorific output.

In a study of Kofyar farm labour, Stone *et al.* (1990; Stone, 1997) employed unweighted labour units. It is true that households cannot change the age and sex structure of their workforce, which forms the overriding constraint on their activities. However, decisions about the allocation of individuals' work, within a multi-sectoral household economy and a culture which lays down rules about male and female participation in farm work, do take account of the efficiency of individuals. Of course, a crude and rather arbitrary weighting system is an imperfect measure, but (we believe) better than none.

2 How much energy does a day's work consume?

A second question concerns the variable *energy* input of an hour's or a day's labour on different tasks, especially when farming and non-farming tasks are compared. Weeding under the sun at the peak of the growing period, for example, is constrained by physical energy, whereas some far less energetic business tasks (for example, selling goods), are constrained more by available time. Domestic work varies between the high energy demands of drawing and carrying water (for example) and a physically less demanding task such as shelling cowpeas. Therefore, to use time as a unit of measurement of labour, while preferable to assigning notional monetary values, remains an imperfect strategy.

3 What is the appropriate measure of a worker's time?

A third question concerns the choice of *time unit* for the measurement of labour inputs. Many agricultural economists assume that the western concept of an hour is appropriate. Why? The working day is normally broken into two, separated by afternoon prayers (at about 2 p.m.): morning (from sunrise until heat or exhaustion compels a rest, which can be before midday in severe conditions) and afternoon (from some time after prayers until sunset). Of course, the actual hours worked vary with the urgency of the task. But the task is perceived in terms of mornings and afternoons, and the intensity of effort reflects a worker's intentions with regard to completion. There are six full days available for such labour and on the seventh (Friday), morning work stops a little earlier in order to prepare for mosque, and recommences (if at all) rather later than on the other days. Such variability cannot be recognised in standardised monitoring. We recorded work by morning and afternoon, assigning different weights to each, later aggregating to days.[4]

4 How long is the farming season?

The fourth question is the choice of *time period* for aggregating gross annual labour inputs. While the major farming tasks (planting, weeding and harvesting) fall within the observable growing period and can be described in terms of a start, middle and end, other supporting tasks (land clearance and fertilisation in the early part of the year and harvesting residues in the later) may be extended well outside the growing period and are not subject to climatic control; that is, they can be done, within limits, whenever it suits a farmer to do them, and at high or lower levels of energy input. This problem – the diffuse boundaries of the farming year – affects any seasonal farming calendar, and parallel considerations affect the use of labour in livestock activities. Since the debate about labour intensification revolves around labour inputs per hectare within given periods of time, this problem needs to be made explicit. Labour inputs before or after the main growing period may significantly affect the totals for the year. But to collect farm labour data throughout the year (the obvious solution) both overlooks the low effort of some out-of-season work and is subject to some practical difficulties, such as tracing and defining 'work' done in small or casual bouts.

5 Comparing like with like

The fifth difficulty is that of *compatibility* between ecological regions where the length of the growing season, or the time available to do farming tasks, varies. This is as true of the rainfed farming season as a whole as it is of individual tasks, such as the time available to complete the weeding of millet. Exactly the same pitfall awaits inter-year comparisons made in the same place. Unless this difficulty is borne in mind, there is a risk of measuring gross annual farm labour in seasons of different length and crediting the longer with a higher level of intensification.

To deal with the fourth and fifth questions, we will argue (in Chapter 6) for the use of *peak* labour intensity as an alternative indicator of labour intensification.

Components of the system

The household (and village) economy is run as a whole, and the interactions between its components (such as crop production, livestock management, marketing, food processing, resource access, migration) involve decisions based on such considerations as the opportunity costs of labour time, cultural practices regarding the division of labour between the sexes, food sufficiency at the household level and many others.

The approach begins, therefore, with inventories of technical options: (1) the use of biological resources; (2) soil fertility management, and (3) livelihood management. The impact of selection on cultivated biodiversity is shown in an investigation of the management and genetic diversity of pearl millet; the impact of land use allocation and incremental investments of farm labour is shown in an analysis of land use and vegetation change over several decades (Chapter 5). The effectiveness of soil fertility management is evaluated by means of a staged analysis of farming intensity (Chapter 6), which quantifies the role of additional labour in what we call 'indigenous intensification'. Diversification out of farming is systematically described (Chapter 7), and the place of farming in household livelihoods is investigated in relation to women and children (Chapter 8).

By means of this three-pronged approach to biological, soil and livelihood resources, we aim to portray labour management as the focus of the diversified household economy.

3

FOUR COMMUNITIES, FOUR SYSTEMS

Landscapes and communities

Seen from the air, the Sahel forms a seemingly endless tapestry of cultivated fields, dispersed among natural woodland (or grassland), for the most part bereft of rivers or lakes, hills or mountains, and marching across the time zones from Cape Verde in the west to the River Nile in the east. A closer look reveals, however, a persistent patterning around the clusters of habitations that make up its countless villages, each surrounded by a ring of fields, which merge together as the villages become larger or closer. The camps of mobile livestock keepers, scattered in the bush, cannot usually be seen. This vast zone, which would take six or seven hours to fly across, has been domesticated over many centuries, using hand technologies.

Yet for the most part this is not a settlement frontier based, like the plains of North America or Australia, on the individualism of the pioneer homestead. Households were linked by kinship, by patronage, by religion or even by slavery, as well as by voluntary association, into communities whose political integrity gave expression to the need to regulate access to the natural resources, share labour, and define the patterns of social relations. These functions of community remain important even as the tentacles of the global market penetrate ever deeper and more and more people participate, as individuals, in commercial forms of exchange, and assert rights, as individuals, over natural resources. Hence it still makes sense to speak of a 'village system', in which members (by right of birth or settlement) negotiate among themselves about rights of access to, and benefits from, natural resources. Indeed, in several countries of francophone West Africa, current policies of 'decentralisation' are attempting to recover this integrity in the concept of *gestion du terroir*.

In each of these four places, communities have come into existence and evolved their own specific resolutions to the Sahelian constraints which we identified in Chapter 2.

Tumbau (Yusuf, 1996; Harris, 1996)

The Kano Close-Settled Zone has attracted considerable attention from researchers, who have all been impressed with the capacity of its resourceful Hausa population to support itself at rural population densities well above the average for the drylands (Grove, 1961; Mortimore and Wilson, 1965; Hill, 1977; Mortimore, 1993a, b; Harris, 1996). The intensive smallholder farming of this ring of densely populated land around Kano city, and its park-like, almost manicured appearance was noted by nineteenth-century visitors such as Barth (1857). Not only does farming support a dense human population, but there are also more domestic animals per hectare (Hendy, 1977; Bourn and Wint, 1994).

The village of Tumbau (in Gezawa Local Government Area, Figure 3.1) is approached from Kano, to whose market (and others) it enjoys good access across 50 km of flat or very gently undulating upland about 500 m above the sea. The orange or brown sandy soils betray the field of dunes that covered this region during the last Quaternary desert transgression (Nichol, 1991). However, their infertile origin is belied by the luxuriant growth of sorghum and millet which, standing up to three metres high in a wet year, shade the low-lying, spreading cowpea and groundnut. Small square fields of less than a hectare, with their boundaries marked out by hedges and clumps of perennial grasses, contain a scatter of mature farm trees. Of various species, most of them

Plate 3.1 Tumbau: maintaining fertility with heaps of manure (*taki*) on the fields of the Kano Close-Settled Zone.

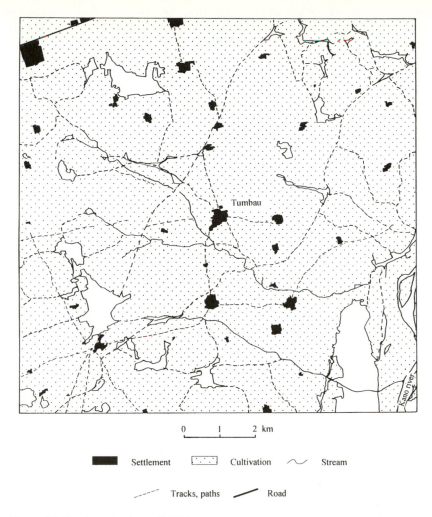

Figure 3.1 Tumbau: land use (1981).

indigenous to the natural vegetation, they break the force of the wind in the long desiccating dry season, when the soil stands bereft of cover and the animals must live on stored fodder. At such times it is possible to observe how cultivation has replaced all but a few patches of the natural vegetation (and those usually degraded soils, or tracks for moving livestock between the fields), under the intense pressure to find places to grow food for an ever-growing population.

The rural Kanawa lived in houses dispersed among their fields, until a late colonial programme of voluntary resettlement created large, gridded villages spaced at short intervals across the landscape; however, many still prefer to reside in their fields. The house, surrounded by a thatched fence, is the locus of domestic activity, of the storage granaries, the corral for domestic animals during the growing season, and the toolshed for farming and for other off-farm activities which are pursued by Hausa women and men when their farming or domestic work allows them time. Gradually, cement and metal sheets are replacing the mud walls and thatched roofs as, *naira* by *naira*, households struggle to save and invest in their future. Yet, each generation, the farms are subdivided between the male heirs, and the task seems more challenging. Is this a pathway of involution from which there is no escape, or a model of thrift and hard work for others to emulate?

The Kano Close-Settled Zone has 'an example of a farming system which has reached the point in the intensification process at which all land is under cultivation, all palatable crop residues are used as fodder, and trees are conserved' (Harris, 1996: 13). Soil fertility is maintained by labour-intensive management involving the integration of livestock and crops (Yusuf, 1996). Long before inorganic fertilisers became available, and even now for those who cannot afford them, *taki* (animal manure, often dry-composted with bedding and compound sweepings) is a vital factor in maintaining the physical properties of the soil and in providing chemical inputs (particularly phosphorus). Hence every family aspires to own as many small ruminants as possible. The intercropped legumes are believed to contribute nitrogen by fixation. To suppress competition from weeds, the growing crops are weeded assiduously three or even more times during the growing season, and, to make the best use of the available plant biomass, everything is harvested and stored or fed to animals – grain, residues, boundary plants, and the weeds themselves. Trees are browsed, or their foliage cut for fodder. Fuelwood too is harvested from dead wood and cut branches – but the trees are privately owned, so they are protected for the future.

The main subsistence crops are sorghum and millet (both early and late varieties). Groundnuts are sold, and cowpeas (again, both early and late varieties) mostly so. However, a distinction between 'food' and 'cash' crops is unsustainable, for the farmers may sell any of their crops at any time. The groundnut, which made Kano famous in the world markets, disappeared after the Sahel drought when rosette disease was widespread in northern Nigeria; however, after a long period of depressed fortunes, it has recovered recently with the introduction of improved, higher value lines. The residues from sold crops are still valuable as fodder, thereby re-entering the nutrient cycle, though residues may also be sold locally. About one-third of farmers grow cassava, which, on account of its long growing cycle (about a year), is grown in enclosures protected from wandering livestock by impressive earth ramparts surmounted by thorn fences – a major investment of labour. Peppers, bambara nut, benniseed, okra, sweet potatoes and tomatoes are also grown, as is rice in a few waterlogged pockets. Not far away, on the flood plain of the Kano River, irrigated dry season cultivation of market crops (vegetables, tomatoes and others), perennials (sugar cane) and wet season rice are possible; only a handful of Tumbau farmers, however, have found the investment funds and travel time necessary to enter this sector.

The cropping systems are highly complex, and diverse in terms of both pattern and intensity. The general pattern is intensive mixed cropping, with at least two major crops and two or three minor ones, all growing in a single plot of a fraction of a hectare. The chosen pattern is determined partly by the needs and interest of the farmer, but also by the fertility status of the field (which is the outcome of the farmer's earlier management). There are, however, three main categories: 'grain main' fields (*gonar hatsi*), 'groundnut main' fields (*gonar gyada*) and those fields in which a portion is devoted to cassava, okra, benniseed or other minor crops (Yusuf, 1996).

Crop rotation is a little studied aspect of local farming practice. Most common is a rotation between 'grain main' mixtures (sorghum and millet) and 'groundnut main', as a strategy to maximise yield and conserve fertility. Rotation may be after either one year or two years under a mixture. It is considered that 'grain main' mixtures can be rotated with 'groundnut main' mixtures with no regard for the fact that early millet or sorghum was grown as an intercrop with the groundnuts in the previous year. Thus a typical rotation might be as follows: first year, groundnut mixed with either sorghum, early millet or cowpeas; second year, early millet mixed with cowpeas, or sorghum with cowpeas.

All this was achieved by means of hand technology, using a small range of tools including the ridging hoe (*galma*), weeding hoe (*fartanya*), cutlass and knives for harvesting. The introduction of the ox-drawn plough was delayed until several decades after it had made a major impact in less densely populated areas. It was assumed that the reason for this was the small size of the holdings, on which the investment would be uneconomic. But today, the majority of farms in Tumbau are ploughed, either before planting (in ridges about a metre apart), or shortly after (for weed control and ridge building). Plough owners contract their teams to farmers in the hands of experienced (hired) plough-boys, whose dexterity at ages as young as 10 years as they manoeuvre the heavy instrument between rows of standing grain or turn on a sixpence at the boundary of the tiny field, can be amazing to behold.

The farming families of Tumbau have thus devised their own resolution of the challenges of the Sahelian environment: taking full advantage of higher and more reliable rainfall than the average for the Sahel, as well as their increasing numbers and their own propensity to work, they have intensified *in situ*, integrating crop with animal husbandry, fertilising the soft, sandy soils, and maximising productivity with a demanding schedule of intricate farming operations which runs throughout the growing period. During the dry season they exploit opportunities for some income diversification through the markets of the region. In addition, they grow trees (some of them leguminous) at densities of between twelve and fifteen per hectare, in a formation called 'farmed parkland' (Cline-Cole *et al.*, 1990).

Dagaceri (Mohammed, 1996; Mortimore, 1989)

Dagaceri Karaguwa (in Birniwa Local Government Area, Jigawa State; see Figure 3.2) is very different from Tumbau. About 150 km north-east, and 150–200 mm drier in most years, it stands on hummocky dunes from 5 to 10 metres above the surrounding country, and 350 m above sea level. It is approached, from the 'coal tar' road that links it with Nguru and Kano, along a sand track that crosses alternate bands of degraded transitional woodland and open cultivated fields, in which the villages stand out on account of their dense clusters of shade trees (especially the neem, *Azadirachta indica*). The contours of the old dunefield are obvious here, especially in the late dry season as the wind blows the sandy topsoil from the empty fields into the grass or shrub boundaries. 'Woodland' is a misnomer for the natural vegetation, which occupies less than half the area and is, in fact, shrub grassland dominated by *Boscia senegalensis* and *Guiera senegalensis*, with patches of dum palms and dum scrub (*Hyphaene thebaica*). There are as many – if not more – trees to be found on the farmland, where massive clumps of *Adansonia digitata* (baobab) dominate a sparse parkland of mature economic trees. Occasional pools occupy the former interdunes for a few weeks after the rains; otherwise, there is no surface drainage.

Three ethnic communities have come, in relatively recent times, to share this habitat, which was formerly a part of Kano State. FulBe groups claim to have used the area first, and a family normally lives in the small clustered hamlet

Plate 3.2 Dagaceri: maintaining fertility with restorative fallows.

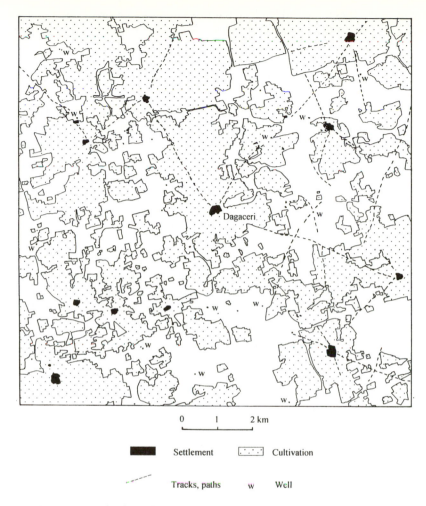

Figure 3.2 Dagaceri: land use (1981).

(*ruga*) with a large farm and a mixed herd of cattle and small ruminants. Some of them practise transhumant grazing with the cattle, taking some or all of them 20–50 km south to the Hadejia River plains in the late dry season. Contesting their claim to prior settlement, however, are Manga farming communities (related to Kanuri) who live in nucleated villages surrounded by rings of permanent and fallowed fields. A *modus vivendi* between the two is recognised in a system of 'grazing reserves' – essentially FulBe territories – on which Manga farmers may not encroach. The third ethnic community, which is specific to Dagaceri, is of Hausa migrants from Damagaram in Niger, who, during the past 40 years, have been accommodated within the Manga village and farm allocating

41

system. This community, which received fresh impulses in the drought cycles of the 1970s and 1980s, is one of *almajirai* or Quranic scholars. The three communities interact constantly at economic, religious and cultural levels, and the Hausa language is the most common medium of exchange.

The village, which had a population of about 770 in 1996, is set out on a grid plan with the Chief's house at the east end according to Kanuri practice. There are three mosques, reflecting the size and diversity of the village. As in Tumbau, there has been incremental investment in more permanent house structures during the past 30 years, as well as other evidence of successful private accumulation by individuals; however, most of these funds come not from agriculture but from incomes obtained outside the village by means of a substantial commitment to dry season migration or *cin rani*. Two open wells have recently been supplemented by a borehole fitted with a hand pump. Unlike Tumbau, Dagaceri keeps its granaries in clusters on the periphery of the village, close to common threshing floors and away from risk of fire.

The innermost fields are cultivated every year, and fertilised using the manure of small ruminants, trash from the village middens and household refuse. But there can be no question of maintaining inputs at levels comparable to those of Tumbau, for families cultivate nearly five times as much land per capita (1.6 ha compared with 0.35). Much of this land is fallowed for a few years after five or six years of cultivation. These fallows carry grass, herb and shrub vegetation. The vigorous regeneration of *Hyphaene thebaica* on both fallows and cultivated fields is noticeable, and only declines on permanent fields. The weeding labour required for these large holdings is considerable. It is common to see fields (or parts of fields) abandoned to weeds late in the season, either because of insufficient labour, or because poor rainfall has affected the crops.

Dagaceri experiences a significantly greater risk of crop failure than Tumbau. The most preferred, as well as the most suitable rainfed crop in such an environment are several landraces of pearl millet, which mature in 90 days or less. The first to be planted it is also first to be harvested to restock the depleted granaries. Millet alone, however, would neither replenish the soil nor supply the diversity desirable to minimise risk. With regard to the first, cowpea is normally grown in mixtures with the millet; while to diversify, a large number of sorghum lines are used. This crop can use residual soil moisture to good effect, and even if planted quite late, as it is photoperiodic, it will still produce a head. Millet is the major source of food, although sorghum is also important. There are seven different landraces of millet and seventeen of sorghum. In the 1960s and early 1970s groundnuts were grown by almost everyone, but after 1975 (on account of drought and rosette disease) groundnut cultivation almost disappeared. The other important crop which deserves mention is the 'cow melon', known locally as *guna* (*Citrullus lanatus*). *Guna* is grown for its oil-bearing seeds, rich in fat and protein. It re-emerged from obscurity as a market crop after 1972, owing to an increasing urban demand (Mortimore, 1989:

106). It is planted towards the end of the rainy season to grow under residual moisture, and so can be planted in millet fields after the harvest has been lifted.

The groundnut was, and *guna* is now, grown exclusively for sale, and cowpea grain is mostly sold (while the residues are, as in Tumbau, valued for fodder). But millet and sorghum may also be sold, and at most times of the year, as surveys in the village show; such sales reflect both farmers' contingent needs for cash and their attempts to profit from constantly changing prices. Benniseed is another crop grown for market, whose recent revival in Kano and Jigawa States is responding to price improvements. However, it is very susceptible to grasshopper attack in Dagaceri.

The major cropping patterns identified in 'millet main' farms are millet with sorghum minor (*gonar hatsi mai sarnon dawa*), millet intercropped with cowpea (*gonar hatsi mai sarnon wake*), and sorghum intercropped with cowpea (*gonar dawa sarnon wake*). It is rare to see any crop growing alone, unless the intercrop has failed. Stands of the cereal grains (which contain from five to ten plants after thinning) are spaced about twice as widely as in Tumbau, and intercrops even wider; this practice optimises available soil moisture and nutrients, but of course increases the area that has to be weeded. Cowpeas are vulnerable to drought and to insect pests, so it is not known how beneficial their nitrogen-fixing activity is for concurrent or succeeding crops.

In Dagaceri, the coarse sandy soils, unreliable rainfall and large cultivated areas call for technologies adapted to extensive rather than intensive strategies. As cultivated holdings were enlarged in the 1950s (partly in response to falling yields, partly in order to grow more groundnuts), the *ashasha* or long-handled hoe (called *iler* in French literature) was imported from further north and swiftly adopted by Dagaceri farmers of all ethnicities. In the 1960s, better-off households gained access to ox-ploughs; but in the Sahel drought, all the ox-teams were broken up by sale or mortality and had to be slowly rebuilt during the 1980s. As in Tumbau, the plough not only saves labour but changes planting practice, from planting in the flat (with the long-handled planting hoe, *sungumi*) to planting in ridges, which, as no hoe is required, is easier and faster. However, as waiting for the plough may cause delay, and such delays are hazardous in the variable rainfall regime of Dagaceri, many fields are still planted in the flat, either immediately after the first rain or in anticipation – a carefully judged risk – before it. A few farmers make use of tractors hired from Birniwa (12 km away).

We can see, therefore, that the key elements of the Dagaceri farming system, by contrast with those of Tumbau, are extensive land use and diverse and flexible cropping practices which can respond to variable rainfall and big changes in constraints and opportunities from year to year. We also observe the partial separation of specialist animal husbandry (in the hands of FulBe) from farming (though Manga and Hausa also keep some livestock). Finally, we also observe a strong commitment to earning incomes away from the village, a commitment which has a long history (potash trading, groundnut exporting) and which tends to intensify when crop yields fall.

Futchimiram (Chiroma, 1996)

In what was formerly the north-western part of Borno State (now Yobe), known to early European travellers as the 'Great Forest of Bornu', extensive areas of wooded grassland still survive, in which the cultivated fields of the Badowoi (who are related to Kanuri and Tubu) form enclaves amounting to only one-fifth of the area (Figure 3.3). These enclaves are managed under a system of shifting cultivation which shows no evidence yet of coming by the terminal stress so often predicted by population growth. Perhaps this is because the population has not grown rapidly (local data are not available), perhaps because critical thresholds have not yet been reached. Whatever the reason, the cultivated fraction has not changed in the area around Futchimiram (which is in Geidam Local Government Area of Yobe State) since 1957 (Turner, 1997).

On this seemingly endless, flat or very gently undulating plain about 300 m above sea level, the savanna woodland is a parkland of mostly mature trees, mainly thorny, drought-tolerant species (*Balanites aegyptiaca*, *Faidherbia albida* and *Acacia*). The grassland, which forms a luxuriant growth about one metre tall after heavy rain, and dies off in the dry season leaving a thin, desiccated (but

Plate 3.3 Futchimiram: the Badowoi parklands in the early dry season.

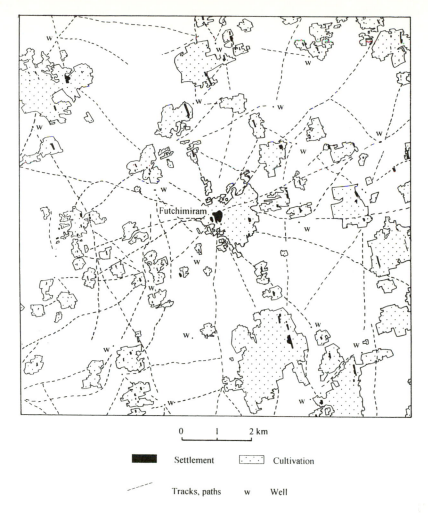

Figure 3.3 Futchimiram: land use (1990).

still edible) cover, is dominated by the annuals *Cenchrus biflorus, Dactylectenum aegyptiacum* and *Digitaria debilis*. Although the soils are very coarse sands, the former dunes and interdunes are less visible than in Dagaceri. There are a few small surface pools during the rains. Groundwater is deeper than in Tumbau or Dagaceri, reached by hand-dug wells at a depth of 30–40 m. The natural ecosystem provides excellent year-round grazing, except when affected by drought. Village fields are opened in blocks, and fenced with thorny branches cut from pollarded trees, in order to keep free-grazing livestock away from the crops.

The Badowoi keep cattle, along with small ruminants and horses, which are used for riding by everyone (women included) and not only by persons of status. They also farm millet. It is not known for how long they have combined the two activities, but their cropping system is not a mere adjunct of animal husbandry. Their hamlets consist of lines of houses set out from north to south, in which the doorways open to the west. In front of these doorways, each household farms beside its neighbours, creating long strip fields reaching to the western boundary of the cultivated block. Behind the house (on its east side), the livestock – particularly small ruminants – are penned, when not corralled on the cultivated fields or out in the rangeland. The small hamlet (between ten and 50 households) used to move every two or three decades when new land was opened up, accompanied by fission and fusion as individual households would move around; recently however, this type of movement is stabilising as land and sites become more scarce.

Apart from corralling animals on the fields, manuring has become an integral part of the farming system, and many farmers collect manure from the pens behind the house and transport it to their fields nearby. More distant fields are dependent on long fallowing for fertility restoration. Surprisingly (for the visitor), almost every hectare of woodland has been cultivated at some time in the last 20–40 years. Plant indicators of soil fertility are cited before authority is given, by the Chief, to open new fields. Perhaps because of such regulation, this system, with 1.4 ha per capita, does not use land on as large a scale as that of Dagaceri. Like the Manga, however, the Badowoi frequently over-plant and have to abandon parts of their farms to weeds. Some complain of declining fertility on the inner fields which have now been cultivated (in the study hamlet) for about 40 years with manure. This manure, it may be noted, contains nutrients transferred from the long fallows where the animals graze.

With even less rainfall than Dagaceri, only three landraces of pearl millet are grown, though it remains the dominant crop because of its tolerance to drought and its ability to grow on infertile soils. Sorghum is widespread but very secondary in quantitative terms; it may be primarily produced as livestock feed. Groundnut (four varieties) is the most important crop after millet. The introduction of an early maturing (upright bunching) variety, about ten or fifteen years ago made the higher yielding (spreading) variety less popular. This may have been a response to lower rainfall expectations. There is no recollection, in the study hamlet, of rosette disease; it seems that groundnut has been grown and sold continuously since it was introduced to northern Bornu via buying agents in the 1950s. Cowpea (five varieties) suffers from pest and disease problems, and frequently fails; a recent outbreak of cutworms in the study hamlet discouraged many from growing the crop, and few now devote much land to it, although there is much interest in possible improved lines. There are few subsidiary crops. This rather limited repertoire of cultivars may reflect the divided priorities of the Badowoi between livestock and crop production as much as the constraints imposed by their environment.

There is, however, no doubting the importance of selling crop output.

Groundnuts contribute more to the total farm income than any other single crop. The shelled kernels can be bagged and sold in the market, the oil from the kernels can be extracted and sold as vegetable oil to be used in frying or in stew making, the leftovers from the processed kernels made into groundnut cake and eaten or sold, and the crop residue itself may be either sold or reserved for use as livestock fodder. Most women who own farms independently from their husbands devote them to groundnuts. Cowpeas are the second cash crop, at least in years when there is a worthwhile output, but they are also eaten at home, and the residue sold as fodder or fed to the family's livestock (particularly to small ruminants) during the dry season.

All crops are planted with virtually no prior land preparation. The loose, sandy nature of the soils makes ridges shortlived, and the labour implications of making ridges using hand hoes (the *ashasha* is not used here, but instead the older, slower though more thorough *fartanya* type) also discourage many farmers from planting on ridges, so crops are generally grown on the flat. At the time when this study was conducted, no ox-ploughs had been introduced to the study hamlet, and some thought the oxen would be more useful in drawing carts. In two additional ways, cropping practices differ markedly from those of Dagaceri. First, monocrops are frequently found, both of millet (that is, without a cowpea intercrop) and of groundnut (which is also common in Tumbau). Second, stand densities are significantly higher. This is to reduce weeding work, especially with groundnut, which is assiduously weeded in closely packed stands on portions of fields which are moved around from year to year.

In Futchimiram, the environmental challenge has been resolved through a mixed enterprise farming system including livestock (especially cattle) and the production of a limited range of crops, emphasising production for the market as well as for subsistence. The degree of integration between the two enterpises is quite limited. But it would not be correct, without qualification, to characterise the system as extensive, even though the cultivated fraction is quite small and the Badowoi still practise collective shifting cultivation. For within their fields they sustain small areas under annual cultivation for many years, and also grow groundnut, if not always other crops, at high densities and with much labour.

Kaska (Ibrahim, 1996; Mortimore, 1989)

The last village on our transect lies in marginally drier rainfall conditions than Futchimiram and in a natural landscape (the Manga Grasslands) which is dramatically different from those of the other three villages. During the terminal phase of the last desert transgression, an area about 150 km wide and 100 km from north to south continued as a field of moving dunes. Since then, soil formation has been very limited and, though annual grasses cover most of their slopes, woodland appears not to have colonised them, except for a few individuals of *Faidherbia albida* or *Balanites aegyptiaca*. During the relatively wet

Plate 3.4 Kaska: a *kwari* depression in the Manga Grasslands.

years before 1970, the grassland was dominated by perennial grasses (including *Andropogon gayanus*), according to resource surveys of the time, but since 1977 (if not earlier), it has been dominated by the annual, *Cenchrus biflorus*. Such transitions have been recorded elsewhere in the Sahel. The ecology of the area has thus gradually changed to a more Sahelian type of vegetation (Figure 3.4). At the same time, dunes have been remobilised and in some places up to 20 per cent of the surface is now bare sand.

Kaska, which lies in Yusufari Local Government Area of Yobe State, is close to the border with Niger Republic. The dunefield, which often looks as fresh as if it were formed yesterday, is thinly covered with grass and can easily be remobilised by the wind in dry years; it is called by the Hausa term *tudu*. Situated 350 m above sea level, it is interrupted at irregular intervals by great depressions, often of triangular or hourglass-shaped plan. They are of two kinds.

The first are steep-sided, flat-floored depressions up to 20 m deep with dense woodland that open up suddenly at the unsuspecting traveller's feet. They are bounded by dunes with slopes of up to 35 degrees, bright orange or yellow sands, and wind-formed crests. This type is called the *kwari*. In it may be found a large, usually seasonal pool of sodic or alkaline water called a *tafki*, and an extensive dry floor of white or grey hardpan (Carter, 1994; Alhassan, 1996). This in turn is surrounded by brown soils supporting a dense woodland of *Hyphaene thebaica* (unless it has been cut down) or, rarely, a mixture of trees including *Acacia* species. In the past the *tafkis* retained water for years

48

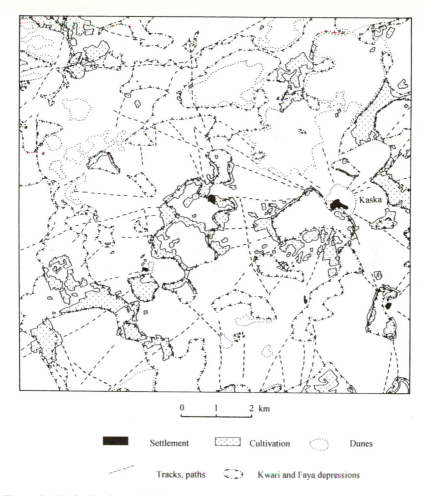

Figure 3.4 Kaska: land use (1990).

(Holmes *et al.*, 1997), and supported aquatic life (fish, hippos and crocodiles). In the dry season when the surface water evaporates, the *tafkis* produce potash (*kanwa*) and salt which is processed locally (Manga: *manguli*). Some *tafkis* have gradually dried up and stopped yielding potash. Only two or three are still perennial, and the aquatic life has disappeared. The moist soil, where it is not too saline or alkaline, supports the cultivation of crops by small-scale irrigation (*garka*, Manga), as water is available a metre or two below the surface.

The second type of depression is shallow, not reaching down as far as the water-table, with a hummocky surface, sandy grey soils and scattered trees (*Hyphaene thebaica, Commiphora africana, Calatropis procera*). This type is called *faya* (Manga). It appears to represent a former *kwari* infilled with blown sand.

The topography of the area has produced four types of farmland. Around the *tafkis* the soil is black clay, rich in potassium, phosphate and other nutrients. Where the soil is moist and the water-table is shallow, small-scale irrigation is conducted. Wells and dug pits provide the main sources of water for the irrigation. Cassava is the major crop grown in these *garka* plots, which are subdivided into very small parcels among almost every household, and rice, wheat, sugar cane, maize, onions and various vegetables are grown. Economic trees – date palm, mango and citrus – are also planted, especially close to villages.

Until about fifteen years ago, the *kwari* type was preferred for rainfed cultivation. The addition of sand to the clays produced a good brown soil suitable for millet, sorghum and cowpea. But persistent drought has made the *kwari* areas unpopular, as these soils are difficult to manage when dry, and moisture becomes inaccessible to the roots of the crops. Thus most of the rainfed *kwari* farms are undergoing long fallow, and some have been abandoned to bush dominated by young dum palm.

The intermediate *faya* depressions contain the greater part of rainfed farmland, where cultivation takes place on an annual basis, with some manuring or with short fallows. Millet, sorghum, cowpea and *guna* are grown. It is inferred from the distribution visible on air photographs that soil moisture or nutrient conditions in these soils were superior to those of the *tudu* dunefield, which were ignored until the drought cycle of the early 1980s. Since 1980, however, a surprising development has been the rapid extension of millet and cowpea mixtures on to the *tudu*. Graphically clear from recent air photography, this is a paradoxical shift under conditions of persistent drought, away from the lowlands and on to the uplands, also reported from adjacent areas of Niger Republic (Reenberg and Fog, 1995).

The Manga Grasslands are inhabited by settled Manga farming communities, whose organisation and settlements are identical with those of Dagaceri, and sedentary FulBe households who also farm in the *faya* (but not in the *kwari*). Owing to their large livestock holdings, the FulBe have more manure to maintain their *faya* farms under cultivation for as many as 50 years in succession. But their major interest is their cattle, and it is notable that, notwithstanding the decline in rainfall and increase in grassland degradation, they normally stay in the area throughout the year. The Manga villages perch on dune tops adjacent to their *kwari* or *faya* farmlands, accompanied by a few shade trees and clusters of granaries. The FulBe settlements are smaller, and scattered. Territorial boundaries between the two communities are conscientiously observed.

Although lacking the large livestock holdings of the FulBe, the Manga have a more diversified economy. Irrigated and rainfed farming have already been described, and the irrigated sector is strongly oriented to the market, even though the road system is non-existent and transport costs are very high.

What sustains the transport system, however, is the exploitation of the potash deposits in the *tafkis*, over which the Manga have local control through the payment of rents to the district aristocracy, an ancient expression of political control (Lovejoy, 1986). Every year, trucks struggle through the sand to reach the stockpiles of potash accumulated by the shores of the *tafkis*, thence to send it, via a marketing network, all over Nigeria. Before the trucks, transport was by camel (a trade which involved some Manga men from Dagaceri, more than 50 km away).

In this complex situation, a resolution of the Sahelian challenge is being sought even as the environment changes under the influence of declining rainfall. As this brief description shows, it would be unwise to predict the form which such adaptation will take. But one characteristic of desert economies is already apparent: the strategic role of markets and of resources located at points (some authors have not hesitated to call the *kwari* depressions 'oases': e.g. Carter, 1994). Diversification, the dominant theme of this economic system, is functionally related to the opportunities provided by physical geography. In this respect, the inhabitants of Kaska, surrounded as they are by obvious evidence of land and groundwater degradation, may be better off than those of, say, Dagaceri, where the options are so much more restricted.

4

NEGOTIATING THE RAIN

Prologue

After the spring equinox of 21 March, Sahelian days slowly lengthen, and the midday heat rises remorselessly. The long dry season (*rani*) passes into the hot season (*bazara*) of local weather lore, and the daily maximum temperatures reach a climax, in May or June, of 40–43°C, depending on latitude. The cooling Harmattan haze has, by then, long since gone. The land stands bare of growing grass or herbage, the surface pools are but cracked hemispheres of hardened clay, the stumps of last year's crops stand lifeless on the field ridges, and the herds and flocks scavenge the field detritus or burnt pastures for the last morsels of edible fodder, their bony bodies testifying to the season of starvation. Hungry it is too for households which have failed to produce a year's food or income. This is the time when the granaries, standing in little groups on the village perimeters, become empty and fall into disrepair. For some they have long been empty, and food has had to be purchased, often selling animals and other valued assets in order to do so. For other, more fortunate ones, a surplus of grain may still remain, putting them into a position of considerable social advantage. Meanwhile, young men return from their dry season migrations (searching for work or trade), perhaps bringing cash to subsidise the livelihoods of relatives, or even invest in a plough, or fertiliser or seed. But for all, this is a time of discomfort, when the men and boys sleep in the street at night to find a little breeze, when the frequency or abundance of the daily meals declines, and sums are done to ration out what remains until the expected harvest in three or four months' time. It is also a time of enforced idleness, when nobody knows when the rain will begin and everyone fit to work must wait, gathering their energy under the shade trees by day, knowing that the year offers but one subsistence opportunity, a short race lasting a few months, whose start is missed at one's peril.

But the new foliage and flowers on the deciduous trees and shrubs, responsive to the lengthening days even while the conditions for any sort of life seem to grow daily less propitious, offer the promise of change. Eventually the fitful airs give way to a stillness, and the seasonal turnabout in wind direction occurs – not firmly or conclusively, but hesitantly, with many changes of heart. This marks the passing northward of the inter-tropical convergence, bringing wet (south-westerly) air from the ocean in place of the arid (north-easterly) desert wind. All at once the sky is clear of haze, the relative humidity jumps from 10 to over 50 per cent, and small clouds form in the afternoons, only to disperse later. Day after unbearable day, the diurnal cycle of heating recurs, sometimes accompanied by traces of rainfall, returning to the stillness of the night. Such meteorological vacillations can be prolonged for days or weeks from April, lasting until June or even well into July, by which time all must fear the spectre of drought, harvest failure and famine. Eventually (and usually), the towering convective clouds and associated upflows of heated surface air precipitate what everybody has been longing for: a cataclysmic frontal storm that, no sooner formed, races away westward at speeds of up to 50 kph, heralded by furious winds or black dust storms, and ultimately bursting in an orgasm of rain. Twenty mm or more can fall in less than half an hour. The water, ponding up in shortlived pools, swiftly penetrates the desiccated soil and disappears from sight, travelling downward through the sandy profile day by day, activating micro-organisms and releasing precious nitrogen. The growing season has begun.

Rainfall in the Sahel

The uncertainty facing Sahelian communities in their struggle to subsist and to sustain their livelihoods is encapsulated in the characteristics of the rainfall. From the ocean to the desert there is an uneven increase in aridity, and a decrease in the length of growing period for plants. This has been shown to be related to the latitude of places, and its variation has been understood mainly in terms of the northward advance and subsequent southward retreat each year of the inter-tropical convergence (ITC: Kowal and Kassam, 1978).

Variability in rainfall between years: drought

Debates about people and environment in the Sahel for the last quarter of the twentieth century have been dominated by the problem of drought. The wet 1950s, the last decade of colonial rule, offered no warning to governments that drought would present such a serious problem after independence. However, from 1968, a succession of years of low rainfall began, which culminated in the Sahel drought of 1972–1974.

Research soon began to dispel the notion that the low rainfall of the early 1970s was unprecedented. Grove (1976), for example, demonstrated that rainfall was lower in the years immediately preceding the First World War in the Sudan Zone of Africa than in the years succeeding the Second World War. In northern Nigeria the years 1912–1914 were dry, and transport on the River Niger was held up by low flows. Rainfall in Bornu was less than half the mean value, and isohyets moved to 150–300 km south of their mean position. The year 1913 was particularly grim, and was followed by widespread starvation. Research has also shown that there was significant climatic variability in the Sahel on time-scales of both hundreds and thousands of years, and therefore that the low rainfall of the 1970s had many precedents (Grove, 1978; Nicholson, 1978). The persistence of low rainfall in the Sahel was recognised in the 1980s (Lamb, 1982; Dennett et al., 1985). Subsequent research has served to confirm this unnerving fact (Goudie, 1995; Nicholson, 1996).

It is now widely recognised that a search for long-term average rainfall can be misleading in sub-Saharan Africa, and especially in the Sahel. The World Meteorological Organisation's data allow rainfall over two 30-year standard periods (1931–1960 and 1961–1990) to be compared in sub-Saharan Africa (Hulme, 1996) and in north-east Nigeria (Hess et al., 1995). These comparisons demonstrate the problem with treating rainfall in the 1950s and 1960s as normal, as it was initially when low rainfall struck in the late 1960s. There was also a reduction in Sahelian rainfall in the agriculturally critical months of June, July and August, with declines in places of 0.4 mm per day (up to 30 per cent) between the two 30-year periods (Hulme, 1996).

During the past three decades, all four of our survey villages have experienced reduced rainfall, in common with the rest of north-east Nigeria. Rainfall trends in this region in 1961–1990 have been analysed by Hess et al. (1995), using records from four synoptic stations, and show an average decline in annual rainfall over the 30 years of 8 mm per year. Studies on a longer time-frame, using lake bed cores from a closed basin (tafki) in the Manga Grasslands (near to Kaska) show that significant climatic oscillations have occurred since at least 5500 BP, including a prolonged drought between 1500 and 1000 years ago (Holmes et al., 1997). Holmes et al. comment that their paleoenvironmental data show that drought has recurred on time-scales varying from centuries to decades throughout the last 1500 years. Droughts in the second half of the twentieth century are therefore far from unique.

Research such as this has progressively demonstrated the flaws in ideas that humans were responsible for drought in the Sahel by clearing vegetation (or allowing their animals to 'overgraze' it). It was argued, at the time of the Sahel drought, that such 'desertification' caused increases in the albedo (or reflectance) values of the land surface, in the amount of atmospheric dust or in evapotranspiration, and might reduce convective rainfall, a process of 'biogeophysical feedback'. (Charney, 1975; MacLeod, 1976; Walker and Rowntree, 1977). In the 1970s, therefore, Sahelian farmers were being portrayed in the scientific literature as unwitting architects of their own misfortune, and ultimately as influencing larger scale global climate change. Confidence in this 'desertification hypothesis' declined, however, as complex patterns of recorded surface temperature variations failed to match simplistic computer model predictions.

The development of increasingly sophisticated general circulation models (GCMs) led to the consideration of the links between rainfall in the Sahel and global sea surface temperatures and the dynamics of the 'global ocean-atmosphere envelope' (Lockwood, 1986; Parker *et al.*, 1987; Glantz, 1992). As 'global climate change' research has burgeoned, the Sahel has been re-positioned as victim of larger scale climatic variations, with the industrialised world's appetite for fossil fuels and profligate discharge of other 'greenhouse gases', the new engine of global climatic change.

The exact nature of the linkages between Sahelian rainfall and sea surface temperatures, and the El Niño–Southern Oscillation are still unclear. It is clear, however, that if rising levels of greenhouse gases (particularly carbon dioxide) bring about warming of the atmosphere, impacts on rainfall patterns in West Africa can be expected. These might be swift if there are significant and rapid changes in Southern Ocean temperatures or ice cover. Anthropogenic impacts on Sahelian climate are therefore possible, not through the actions of African smallholders and pastoralists (as was once dismissively thought), but rather through the profligate use of fossil fuel energy and the production of green-house gas compounds in industrialised countries. Attempts to use models to predict the impact of global climate change on food supplies suggest that increases of carbon dioxide levels according to high estimates would have only a small negative impact on global food supplies (Rosenweig and Parry, 1994). However, they would increase disparities between developed and developing countries, with significant rises in global cereal prices and the population at risk of hunger in the tropics. Farm-level adaptation could do little to offset these risks. If these predictions are reliable (and the assumptions on which they are built provide scope for controversy), the Sahel may share the severe impacts of rising levels of carbon dioxide and climate change. These impacts are in addition to the climatic variation that has been shown to be normal over the last millennium by palaeoenvironmental studies.

However, the significance of global climate change for the Sahel is not yet clear. Models are still not sufficiently developed to predict regional (i.e. sub-global) dimensions of climate change with any accuracy, nor are they sufficiently detailed to predict small-scale temporal variations in rainfall (especially at the start and end of rainy seasons), and changes in the probability distribution of rainfall within and between years. It is on these scales that climate change may bite into the life-chances of the poor in rural West Africa, for the timing of the start and end of the rainy season, and the adequacy of the rains within that season, are the critical driving forces for Sahelian production systems, both for output and for welfare.

The links between rainfall and production are very clear. Broadly speaking, plant growth is possible when the balance between rainfall and potential evapotranspiration (PET) is positive (Barrow, 1987). Soil moisture conditions and runoff make the picture far more complicated of course, but the ecology of the seasonal tropics is driven by the arrival of rainfall, in particular seasons, sufficient to offset high levels of evapotranspiration. In the Sahel that rain falls in a single wet season, whose length is determined by the advance northward of the ITC, and the penetration of moist air northward towards the Sahara. The significance of the rainy season for biological production is neatly demonstrated by studies using remote sensing. Tucker *et al.* (1991) used multispectral imagery from the advanced very high resolution radiometer on NOAA satellites to assess total herbaceous biomass production in the Sahel during the period of drought in the early 1980s. They demonstrated the spatial and temporal variability in herbaceous biomass production in an area of 30,000 km^2. Total estimated production varied from 1093 kg/ha in 1981 to 55 kg/ha in 1984.

Variability in rainfall within years: drought

Changes in rainfall within and between years in the Sahel have a significance not only for the natural grasslands but equally for the domesticated grasses of farmers' fields – millet and sorghum – and other crops. It is rainfall variability that is the dominant environmental risk facing farmers and livestock keepers in the Sahel, and that risk is primarily related to the timing, length and adequacy of the rainy season. Rainfall in the Sahel is episodic, falling in short, intense showers, only one of which might deliver one-tenth or more of the year's rainfall in less than an hour. Intervals between rainfall events can be long, and the amount of rain in each very variable. Thus it is possible to have repeated small rainfalls over several weeks followed by a month without rain. If a farmer plants under these conditions, the death of germinating seeds is a

Plate 4.1 Abandoned millet fields after a drought (November 1973) near Dagaceri.

Plate 4.2 After good rain: a late millet and cowpea intercrop at Tumbau, September 1995.

likely outcome. Equally, the first rains may be heavy, but the farmer cannot tell if the amount that has fallen will provide sufficient soil moisture to bridge the unknown dry period until the next rainfall. The onset of the rains is very gradual:

> The first falls in the season are variable in amount and irregular in frequency; however, once the increase in the amount of precipitation has reached the rate of about one 1 inch per decade (25.4 mm in a 10 day period), the frequency and the amount of precipitation regularly increases per succeeding decade and rainfall becomes reliable and adequate to the extent that there is a surplus over evapotranspiration demands which rapidly replenishes the soil water deficit of the dry soil profile.
>
> (Kowal and Knabe, 1972: 25)

Managing variability is difficult for the farmer (and a reason for the weather lore, debate and indigenous knowledge he or she brings to the task), but it also creates difficulties for the researcher. Hulme (1987) classifies wet season definitions broadly into two: 'absolute' definitions formulated in terms of daily or monthly rainfall totals, and 'relative' definitions. Absolute definitions are simple, for example, defining the wet season as starting in the first month with more than a minimum quantity of rainfall (e.g. >60 mm), or in the first two-day or five-day period with more than a certain rainfall (Stern *et al.*, 1981). Relative approaches include identifying the first month to have more than a stated proportion of annual rainfall, or to differ significantly from the previous month; or they may be based on the reliability of rainfall. Evapotranspiration can also be included as a factor using a water balance model. Agnew (1989) included evapotranspiration in a water balance model of moisture stress for pearl millet. This allows meteorological drought (the impact of a shortfall from normal) to be distinguished from agricultural drought (the impacts of a dry year on, in this case, millet). Hulme (1987) took the beginning of the rainy season to be the first of 120 consecutive days with a positive water balance. Kowal and Knabe (1972) mapped the onset of the rains in Nigeria, using the first ten-day period in which rainfall is equal to or greater than one inch (25.4 mm) and followed by a second ten-day period when rainfall is equal to or exceeds half the evaporative demand.

Page (1994) modelled the agroclimatological implications of reduced average rainfall for northern Nigeria, and of various scenarios of global warming. A two-degree rise in average global temperatures would raise PET by about 6 per cent in the growing season (ibid.: 23–27), causing effective reduction in the length of the growing period.[1] The model shows that an excess or deficit of 20 per cent above or below the mean rainfall for 1961–1990 would have a substantial impact on the length of the growing period. Of the three stations taken to represent northern Nigeria, Nguru is the most affected by a reduction in rainfall, Kano is less so, and Maiduguri is intermediate. Subject to the stated assumptions, the model shows the agroclimatological effect of a continuing decline in Sahelian rainfall in the future. While it is impossible to predict rainfall trends (and a reversal of the declining trend of the last 30 years is as likely as its continuation), the model helps to define the adaptive challenge being faced by agropastoralists.

Rainfall in four villages

The four villages of Tumbau, Dagaceri, Futchimiram and Kaska describe a transect across an ecological transition from semi-arid to arid (according to UNEP, 1992) or from moist to dry semi-arid (according to FAO, 1982; and see Figure 4.1). The critical isohyet in this transition is that for 400 mm mean annual rainfall. As the rainfall declines from south-west to north-east its variability increases, because the number of rain events becomes less and the impact

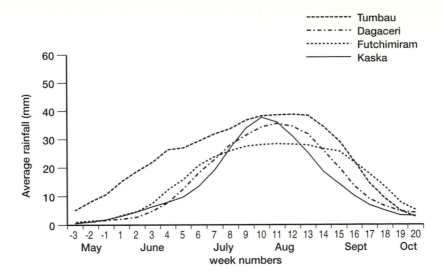

Figure 4.1 Average rainfall in four villages, 1992–1996 (7 week running means).

of one or two missing (or extra) events becomes correspondingly greater. There is a corresponding decrease in the quantity of plant biomass, reflecting the inability of plants to photosynthesise on account of moisture stress and increasing rates of potential evapotranspiration; this trend is shown in vegetation indices derived from NOAA satellite data (UNEP, 1992: 36). The challenge to production systems (farming and livestock keeping) and to livelihoods which is posed by moisture deficits therefore increases along our transect from Tumbau north-eastwards.

Daily rainfall was measured in the villages in the five growing periods of 1992–1996.[2] The average rainfall in the years observed differed considerably among the four villages. It was 571 mm in Tumbau, 360 mm in Dagaceri, 375 mm in Futchimiram and 345 mm in Kaska.[3] These amounts are representative of the reduced levels of Sahelian rainfall that have been normal since the 1970s.

In the graphics accompanying these chapters, village rainfall is presented in seven-day periods, counting from 1 June, and standardised for all stations and for all years. The same periodisation is used for labour-use data presented throughout the book. Figure 4.1 shows the running means for seven-day periods. Over the four years, the average rainfall patterns display a predictable spatial trend from Tumbau to the drier villages: a shortening of the rainy season, and a sharper peak in July to August. Rainfall in Futchimiram, however, failed to peak as expected.

This said, however, what matters to villagers engaging in the risky business of farming or livestock breeding is the variability of the rain, both in time and in

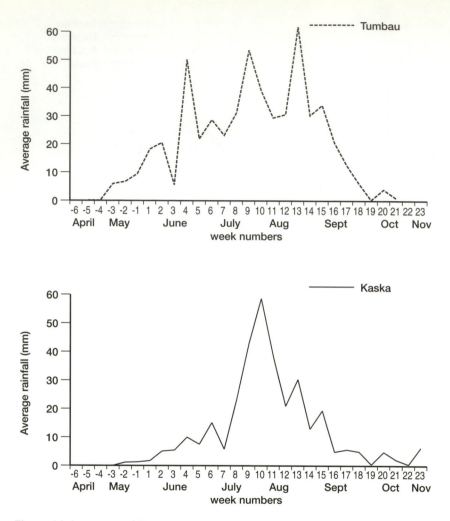

Figure 4.2 Average rainfall in Tumbau (1992, 1993–1996) and Kaska (1992–1996).

space, and its unpredictability. The drier the place, the less its rainfall conforms to the average.

Inevitably, much of the analysis of rainfall in the Sahel in recent decades has drawn on the records of a small number of synoptic meteorological stations with long-term records. Long-term averages provide, at best, only a rough guide to the significance of rainfall variability for farmers and other resource users. Farmers' and livestock keepers' work is exposed to the variability of rainfall both within and between years, and not to average rainfall. The long-term changes in rainfall averages discussed in the previous section are paralleled by variability that occurs from year to year in the actual circumstances of agropastoral producers. To this concern for short-term (inter-annual and intra-annual) rainfall variability can be added the problem of rainfall variability in space. In the Sahel, as elsewhere in semi-arid regions, rainfall is patchy in distribution in space as well as in time. Most farmers live far from synoptic rainfall stations, whose coarse distribution across the region disguises the sharp spatial variability that occurs (Buba, 1992).

Smoothed rainfall curves show that the four villages in Nigeria are broadly representative of the Sahel, but give no indication of short-term variability. Standard measures of variability derived from long-term data cannot be generated from four years' data collected in the villages. These data, however, more accurately express the behaviour of local rainfall from year to year, and within seasons, to which farmers have to adapt.

Just how much the rainfall in a particular year can differ from the expected pattern is shown in Figure 4.2, where the four-year averages may be compared with the smooth patterns for these villages as shown in Figure 4.1. The contrasting humped and peaked configurations of average rainfall in these two villages, shown in Figure 4.2, are now seen to hide very high variability. Rainfall departs from the short-term averages by wide margins in most weeks in both Tumbau and Kaska. There is variability in the start and end of the growing period, in the distribution of rainfall events and dry spells during the season, between different periods in the same year, and between the same period in different years. Given these characteristics, rainfall is perceived, not surprisingly, to be the primary challenge calling for adaptive management, whether by crop or livestock producers.

Working with the rain

The decisions that farmers have to take are primarily concerned with the management of the phenological stages of different crops. The farming year (which runs from the commencement of the rains until the end of the harvest) is dominated by three major tasks: planting, weeding/thinning and harvesting. Land clearing takes place for the most part before the rains, though tasks such as uprooting regenerating shrubs may occur at any time. Fertilising, where it occurs, involves carrying manure to the fields before the rains, and distributing it to the plants during the growing period – sometimes quite late. The ox-plough, which is rapidly gaining over hand cultivation in Tumbau and Dagaceri, is used for ridging before planting (shown in our graphs together with clearing, as it uses relatively little labour), and increasingly (in Tumbau) for early inter-row weeding. The timing of these tasks reflects the growth cycles of the major crops – 90-day early millet, 120-day late millet and sorghum, 70–90-day and 120-day cowpea (the latter only grown in Tumbau), and 120-day groundnut. It also reflects the priority given to pearl millet (mainly early) in all four places.

Planting

The germination of the seed after planting depends on an increase in soil moisture, as during the dry season the topsoils are desiccated. The transition from dry conditions to a sufficiently wet soil in order to stimulate germination and support seedling growth normally occurs abruptly. Farmers may either anticipate this transition by dry planting, or follow it by turning out all available labour (men, women and children) in the two or three days following the first heavy rain. To miss the best of the moisture (and possibly nitrogen) flush imperils later growth and output; but, on the other hand, dry planting risks losing seed, which is sometimes scarce. Farmers usually judiciously mix the two techniques for their first planting of early millet.

Inputs of labour to the field usually peak within 24 hours of the onset of heavy rain. In a hesitant rainy season, from two to five plantings may be necessary before all stands have been successfully germinated and established.[4] Such false starts (cf. Hulme, 1987) place a severe demand on human effort at a difficult time of the year, since farmers may need to make repeated energetic forays on to their land.

Planting by hand (using small-bladed seeding hoes and burying the seed with the foot) usually takes place in unprepared soil. Ox-ploughed ridges, on the other hand, must be prepared before planting, yet they must also await the arrival of rain to soften the soil. This usually delays planting, though better growth later on is considered to compensate for this initial disadvantage. Plough-ridging however can continue for later planted crops until July or even August. Interplanting of longer cycle crops – cowpea, groundnut, sorghum and others – can go on late if a favourable rainfall distribution seems to justify optimism about the residual moisture available to support maturing crops, sorghum in particular. Constructing a dense geometry of interplanted mixtures is one way of controlling weed growth later in the season.

Weeding

Weeding, the second major task in the labour calendar, must begin within a week of planting if the tender crop is to be protected from competition for nutrients and water, and this urgency is greatest when moisture is scarce. Thus a critical decision facing every farmer is when to divert labour from later planting – of longer cycle crops or in outlying fields – to weeding the first plantings of much-needed millet. The sizes and spatial distribution of household fields, which vary between the villages, carry some weight in these decisions, but equally important are the spacing of crop stands which may be twice as high in Tumbau as in Dagaceri or Kaska, and the relative vigour of weed development (which is normally greater with higher rainfall). Owing to these variables, the number of weeding rounds considered necessary differs. In all the villages, two weedings are attempted. In the second, thinning takes place to assure healthy growth of six to eight grain plants per stand. In Tumbau, a third weeding is always done and more may be attempted. But in the drier villages, the larger fields and compressed growing period tend to prevent more than two rounds of weeding, while it is quite common for fields to be abandoned after only a single weeding when family labour is scarce and the farmer cannot afford to hire. Personal misfortune, such as sickness, has profound effects on output through the withdrawal of weeding labour, notwithstanding the likelihood that neighbours may give some assistance.

Weeding technology consists of heavy *galma* hoes in Tumbau, lighter short *fartanya* hoes or long-handled *ashasha* hoes in the other villages, ridging ploughs (which are effective for first and sometimes second weedings as long as the height of the growing crops does not exceed about a metre), and (rarely) ox-drawn weeders. The *ashasha*, used only on the sandy soils of Kaska and Dagaceri, can double the area weeded per day, and the plough is said to increase it by a factor of eight – but while fast between the rows it is ineffective along the rows and has to be supplemented by hand. These ratios enter into farmers' decisions about the disposition of their labour resources.

Harvesting

Harvesting, the third major farm task, begins when early millet is ready for harvest after between 90 and 100 days, when weeding work is suspended – or nearly so – and the 'hungry season' (or famine, if there was one) is at last broken. Premature grain is sometimes picked and eaten before this time, but on a tiny scale. Harvesting – especially of grain – requires much labour and involves many different tasks: cutting the heads; drying, tying into bundles and transporting to granaries; cutting stalks; drying, bundling and removing (unless they are left for animals to graze in the fields); stacking or storing in the branches of trees. After the early millet has been secured, and also early cowpea, there is a return to weeding work on the later crops, but the scale of this task depends on the rainfall at the end of the season, as weed growth is sharply reduced during September. Then comes the harvest of late cowpea, groundnut and sorghum, usually lasting well into November. Late millet has the same cycle as sorghum. Harvesting, therefore, has an extended and variable 'tail' at the end of the season, depending on the crops grown and the rainfall earlier on.

Thus, to interpret the linkages between rain and farm labour we need first to divide labour into these three main categories – other tasks being relatively small in labour terms (Figure 4.3). A comparison of labour inputs by collaborating households in three of the villages will illustrate how many variations there can be on the basic theme.

During a rainy period of 20 weeks in Futchimiram (in 1996), the three labour curves for planting, weeding/thinning and harvesting were distinctly separated. Planting began very abruptly with the first rainfall at the end of May, and

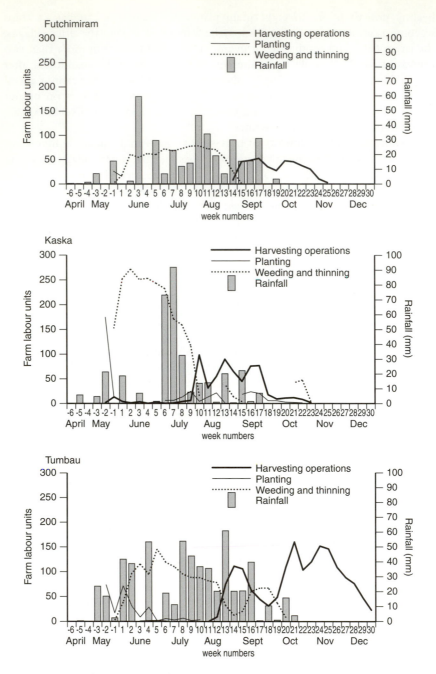

Figure 4.3 Farm labour and rainfall, three villages, 1996. (Labour is shown in weighted units (equivalent to one man-day) per seven-day period, and rainfall in mm per seven-day period.)

declined rapidly. In Futchimiram, early millet, groundnut and cowpea must be planted quickly, given an expectation of a short rainy period and an erratic distribution, giving, in the year in question, an early peak in mid-June. Weeding began in the second week within days of first planting, rapidly attained a high level and maintained it for eleven consecutive weeks. However, as soon as early millet is ready for harvest, labour must be allocated to that. This transition began in the first week of September, and weeding quickly fell off to zero. There were two harvest peaks, clearly distinguishable, the first for millet and the second for groundnut. The farming year came to a conclusion early in November. This pattern represents Sahelian rainfed farming at its simplest.

How representative was this pattern of other places? Although slightly longer (22 weeks), the rainy period at Kaska in 1996 was much more erratic, with eight weeks, two of them consecutive, having negligible amounts of rain. In contrast to Futchimiram, rainfall did not peak until the end of July, having been extremely poor for the first ten weeks of the growing period. The effects of this early drought on the production of biomass are shown, first, in a declining curve for weeding (as weed growth was reduced by drought) and second, in very low labour expenditures on harvesting work. Nevertheless, the configuration of the three curves for major agricultural tasks has the same features as in Futchimiram: an abrupt start and rapid end to planting; a sustained though shorter (six to seven weeks) weeding cycle, a sharp transition from weeding to harvesting, and a multi-peak, though lower and longer harvesting curve.

But a bad harvest of rainfed crops was not the end of the story for Kaska farmers. In the topographical depressions (Chapter 3) there are opportunities for dry season cultivation, and poor results from rainfed farming could be compensated by putting effort into preparing irrigation plots (with peaks in land clearance in September and October, and renewed planting activity) and, later, into weeding work on these plots (November). The continuation of this dry season *garka* cultivation extended beyond the observation period. This demonstrates the labour implications of having an alternative farming option with which to respond to the risks imposed by an erratic rainfall distribution.

In Tumbau, the greater length of the farming year is at once apparent. Although rainfall began (unusually) only a little earlier in 1996 than in other villages, it continued for 23 weeks without a serious break. Neither the short peak of rainfall recorded in Futchimiram nor the erratic distribution of Kaska were apparent in Tumbau, although there was a dry period at a critical point during July.

The configuration of the planting, weeding/thinning and harvesting curves in Tumbau were quite different from those of the other two villages, reflecting its typical experience of a longer rainy period. Several features stand out. First, there were three separate planting peaks in May, early and late June, reflecting the requirements of early planted millet and sorghum, cowpea and groundnut interplanting. Second, planting continued to absorb a significant proportion of labour after weeding had commenced in June. The length of the sustained weeding period was somewhat longer (thirteen weeks) but, as in the other villages, it terminated with the beginning of millet harvesting in August. However, a third major difference was that after the early millet was harvested, weeding climbed to a second peak in September, thereby showing that what brings weeding labour down is the priority of the harvest and not the end of weed growth in this wetter environment. A second transition occurred early in October, followed by a harvesting peak for cowpea, and another for sorghum and other late crops in November. The greater length of the rainy period in Tumbau permits more crop options to be exploited, more weeding to be done and more complex management decisions to be made.

The plotting of temporal patterns of labour use demonstrates how a farming system can be constructed in response to rainfall and how it must be subject to the opportunity costs of competing demands within the farm sector. In particular, the transitions between the major operations of planting, weeding/thinning and harvesting occur when labour constraints are clearly operating, and would reward further study. These three cases show something of the diversity found along an ecological gradient and the links between rainfall, productive opportunities and intensification of effort. They provide a warning against over-simplified conceptions of 'Sahelian farming' or even of a major farming task such as weeding, which can vary with ecological and site factors. In considering the possibilities for technical change, and in particular for labour-saving innovations, there are important management variables to be taken into account as well as capital limitations. This theme will recur later in this book.

Inter-village comparisons therefore show several effects of differences in expected rainfall and in the length of growing periods, on the diversity of crop enterprises, on the conjunction of growth cycles and of bunched labour requirements. The shorter the growing period, the more concentrated the farming effort; so the sharpening rainfall peak, as one proceeds to drier Sahelian regimes, is matched by increasingly peaked labour use.

However, all villages have one feature in common. Given the uncertainty of the start of the rains, all households must commit their labour for several weeks with little productive work in waiting for the commencement of planting. This represents a significant loss in economic terms, for during the long dry season labour is not (as commonly supposed) idle, but engaged in a range of income-earning activities, many of them away from home (Chapter 7).

The diversity increases further if we look at inter-year comparisons. Figure 4.4 shows labour use in Dagaceri in 1994 and 1996. In the first year, rainfall was adequate and well distributed, and this was reflected in high labour inputs in weeding/thinning and in two strong harvest labour peaks. By contrast, 1996 had a drought in July which had the effect of reducing labour inputs (per week) during the weeding period to little more than half the level achieved in the earlier year. This reduced effort was unavailing, and harvest labour – which must be an indicator of output – tailed off at only one-third of its earlier level.

However, Dagaceri has an additional cropping option: the cow melon, or

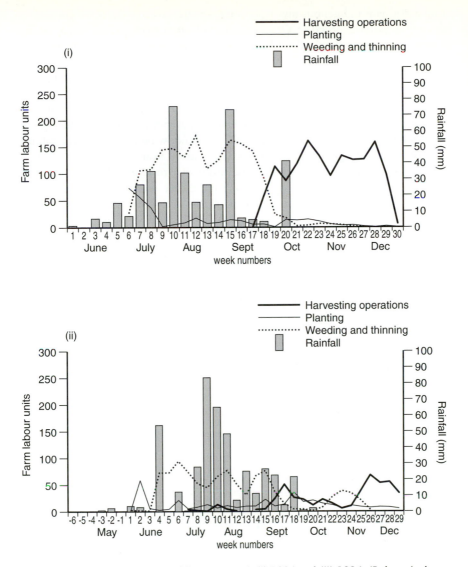

Figure 4.4 Farm labour and rainfall in Dagaceri, (i) 1994 and (ii) 1996. (Labour is shown in weighted units (equivalent to one man-day) per seven-day period, and rainfall in mm per seven-day period.)

guna, which can be planted on millet fields after harvest, and, if succoured by one or two late showers of rain, can continue to grow throughout the dry season, producing a crop of bitter fruit that after time-consuming processing yields an oilseed commanding a high price in urban markets. But this crop is vulnerable to an insect pest which is little understood, as the crop was revived from a dormant indigenous inventory only recently. In 1994, a small amount of planting activity recorded in October reflected *guna* work; but it failed, owing to pests. In 1996 it survived; and there was further labour input on its weeding in November.

Clearly, in 1996 there was – in some sense – a labour *surplus* in Dagaceri. This example shows how mistaken a conventional concept of scarcity – or excess – can be in Sahelian conditions. What matters is not an absolute endowment of resources – in this case labour – but a correct matching of labour, under given technological conditions, not merely with the endowment of land and capital, but also with the opportunities provided in a given year by fickle rainfall. It is not easy to adapt the family labour force from year to year, still less from week to week. It would be true to say that one must have more than enough in dry years in order to maximise the potential of the wet. There is an analogy with opportunistic livestock breeding (Sandford, 1983; Benhke *et al.*, 1993).

Working to the limit

How stressed is family farm labour under differing conditions of rainfall? In order to answer this question, we now shift our analysis from *total* weighted labour used by the collaborating households in each village to the percentage of *available* labour used. Available labour is defined demographically as the sum of household members weighted according to age and sex. It takes no account of the many factors (for example, custom, incapacity, non-farm commitments) which intervene in the mobilisation of this labour for farm tasks. Consequently, the relatively low percentages achieved are not an indicator of inefficiency but rather a measure of the competing demands for and limited capacity of family labour. Nevertheless, the movements of available labour used during the season, and the differences between places, provide suggestive indicators of the pressures that farm tasks exert on the family labour force.[5]

Figure 4.5 shows the percentage of available labour that was used in farming in 1994 and 1995. These were respectively good and poor rainfall years. Three comments are appropriate. First, Tumbau households used their available labour less variably than those of the other (drier) villages, a difference which reflects increasing rainfall variability and risk down through the ecological gradient. Second, Tumbau households used their labour more variably in 1995 – a year of lower and more variable rainfall – than in 1994. This would be expected. Third, in 1994 there was a clear gap between the drier villages and Tumbau, reflecting a categorically greater commitment to farm work in those villages during the period in question. Finally, in 1995 this gap all but disappeared, as labour use in the drier villages fell in response to disappointing crop performance and reduced weed growth.

Such patterns question the appropriateness of introducing a concept of labour use efficiency into Sahelian farming. For many households, the exigencies of the natural environment impose alternating episodes of under- and over-use of

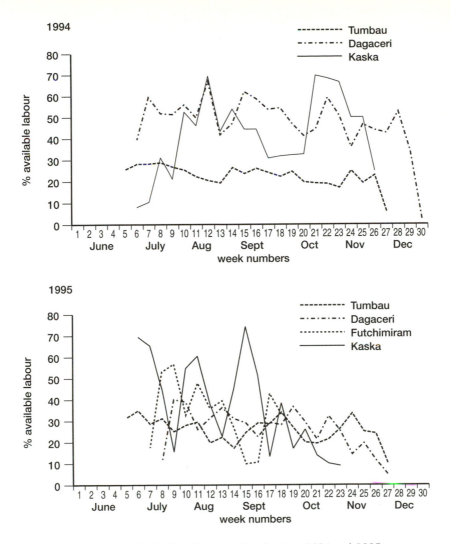

Figure 4.5 Percentage of available labour used in farming, 1994 and 1995.

labour resources. Yet these data cannot express the full extent of the problem. When the labour required for farm tasks exceeds that which is demographically available, the work undone is a labour deficit, perfectly understood by farmers. Such a deficit may be surmountable for individual households by hiring or exchanging labour. For the village system, such household deficits could be taken care of by such arrangements if well distributed in time. But as they tend to coincide (being driven by the rainfall), they represent a source of economic loss in unrealised work.[6] The outcome is poor weeding, inadequate thinning, poor growth, unnecessary harvest losses, reduced yield and output per worker.

Bringing the harvest home

How is all this productive effort brought to a conclusion at the end of the farming year? The harvest is a time of urgent, hard work in any strongly seasonal climate, but in matter of detail it may vary a great deal from place to place. At Futchimiram, with its low rainfall, compressed growing period, extensive farming system and limited crop repertoire, the use of labour for harvesting activities differs substantially from that of Tumbau, with its higher rainfall, more extended growing period, intensive farming system and diversity of crops (Figure 4.6).

Two distinct harvest labour peaks occurred in Futchimiram in 1996: the first was for early millet (in September), and the second for groundnut (in October). No sorghum is grown in this system, and although cowpeas are usually planted, the crop failed in 1996, so there was nothing to harvest. It is noteworthy that groundnut harvesting, which involves several operations (pulling, drying, separating the pods from the haulms, removal for the eventual sale of groundnut, and the storage of the haulms for fodder) used labour – in that year, at least – as much as the harvest of the staple food crop, millet. Groundnut is a major source of income in this system, an importance it has never lost since the 1960s when it was first grown for the market. Since it does not compete directly with grain production for family labour, and its land requirements can easily be met in a relatively land-rich area, it has found a secure place in a farming system that is characterised by a lack of complexity.

Harvest time in Tumbau is very different. A larger number of important crops greatly increases the complexity of harvesting operations. To begin with, there are two millet harvests. Early millet (*gero*), which since the onset of lower rainfall in the early 1970s has been much more important than late millet (*maiwa*) was harvested from the last week of August to the third week of September. This crop matures in 90 days, irrespective of date of planting, being insensitive to day-length; so in a year (such as 1995) when the rain begins later, the harvest may be delayed by three weeks. Early millet tends to coincide with early cowpea, which (also since the 1970s) must be planted, in case the more traditional later varieties fail on account of drought. Early cowpea is interplanted with millet or sorghum, and matures between 70 and 100 days, though much quicker maturity is sometimes achieved under high insolation, further north (50–60 days). It can be seen from the allocation of labour to this crop that competition is resolved in favour of millet, which is the preferred grain staple, even though early cowpea provides a saleable commodity at a time when cash is usually needed. Labour for cowpea harvesting was, as the curves imply, partly deferred.

Next was the harvest of groundnut in October, and this peak can also slide back a week or two when planting has been delayed. Often planted several weeks later than the grains, the groundnut matures in 110–140 days. In this year its harvest did not compete for labour with either early millet and cowpea

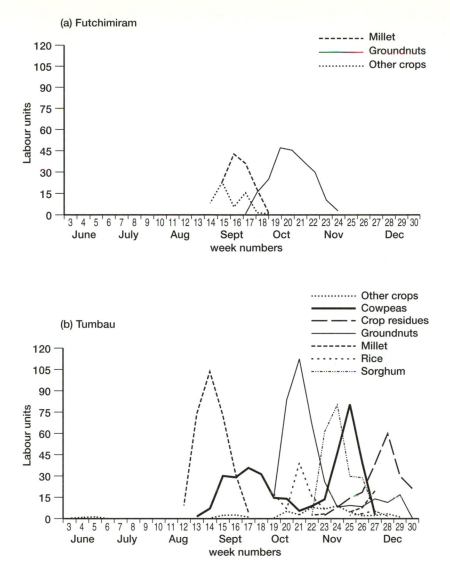

Figure 4.6 Harvesting labour in Futchimiram (a) and Tumbau (b) in 1996. (Labour is shown in weighted units (equivalent to one man-day) per seven-day period, and rainfall in mm per seven-day period.)

(which were already secured) or sorghum and late cowpea (which was still to come). However, a small harvest of upland rice took place at the same time, as there had been enough rain to waterlog certain sites where it is planted as a solo crop. In a drier year, such as 1995, there was less rice, and it was not harvested until November. Both crops are sold. The groundnut, once the foundation of the economy in Kano, was destroyed by rosette disease and persistent drought during the 1970s, but there is a strong demand for it in urban markets (it is no longer exported to the world market), and experiments with new varieties have finally borne fruit in a dramatic resurgence of the crop during the last decade.

The third episode of the prolonged harvest at Tumbau comes when sorghum, late cowpea, and finally late millet are ready. Sorghum is photoperiodic and therefore matures at the beginning of November more or less regardless of its planting date; it is the most predictable element of the cropping system. The growing period thus varies widely, depending on the time of planting, from 110 to over 150 days. Of course, its planting cannot be delayed when the rain comes early because a late drought would damage its growth and yield. Again there is a clash with cowpea, this time the later varieties which mature in 120 days or more, and the curves for 1996 again show some displacement of cowpea by the sorghum harvest. As the two are interplanted, the cowpea is more easily harvested after sorghum stalks have been cut and laid to dry in the furrows. Finally, the late millet is brought in. It has declined to the status of a sideline in today's cropping system, and is often grown along the margins of fields (as it is insensitive to disturbance and low fertility). The harvest of the subsistence and market crops was not completed until the third week of December 1996.

However, this is not the sum total of the harvest activities which demand labour in the Tumbau system. Unlike Futchimiram, Tumbau offers no grazing for livestock during the farming season. So fodder must be 'harvested' from hedgerows, cattle tracks and above all, field weeds, and carried to the house (usually by headload) where it is fed to the animals. In 1995, the labour used for this work (much of it children's labour) actually exceeded that allocated to harvesting any individual crop in ten successive weeks, with two exceptions: the peak weeks of the early millet and groundnut harvests. Furthermore, labour must be allocated to harvesting crop residues, which are too valuable to be left in the fields after the livestock are let out of captivity in November or December. Cowpea and groundnut haulms are stored in the compounds, and grain stalks are either built into stooks on the fields or, since theft has become a problem, hoisted into the branches of privately owned trees to secure them from wandering animals and visiting migrant herds. In 1996, the harvest of crop residues comprised a fourth and final labour peak in December, though when poor rainfall has produced a smaller output of biomass in the system this activity can be moved forward to October, as it was in 1995.

Conclusion

We have shown in this chapter that a statement such as 'rainfall constrains Sahelian agriculture' gravely understates the complexity and diversity of the linkages between rainfall behaviour and the management of farm production. Inter-annual variability, and the compression of the growing period which usually accompanies a reduction in rainfall, impart an opportunistic flavour to the goals and practices of Sahelian agriculture. It is strictly incompatible, either with smallholder farming in equable climatic regimes or with large-scale farming enterprises in which commercial goals, technologies and insurance systems go far to cancel the effects of climatic variability on livelihoods.

Variability, and the unpredictability of this variability, are key constraints that have to be taken into account by smallholders, whether they be better off, 'middle peasants' or very poor. Crop inventories, technological choices and the management of household farm labour all reflect this imperative. A diversity of responses, both between places and between years, calls into question the usefulness of a concept of 'Sahelian agriculture' that is circumscribed by averages, normal distributions and normative prescriptions based thereon.

An analysis of labour inputs through time shows that certain 'inefficiencies' are necessary conditions of success in managing such a difficult environment. First is a necessity of having some 'surplus' farm labour in dry years – when major farm tasks are less demanding of labour – in order to maximise economic output, and control bioproductivity, in wet years – when they demand more. Second is the unavoidability of idle labour during the sometimes long wait for the commencement of the rains, as all hands are needed within 24 hours of the first storm. Third is the well-known mismatch between the employment capability of the system in the dry season and its demands in the growing period, which necessitates in turn an access, through unrestricted spatial mobility, to other sectors of the economy.

The analysis of the use of available labour – in terms of the percentage of demographically defined household labour forces – does not serve usefully as a critique of the 'efficiency' of farm labour (which in view of the foregoing is a meaningless concept) but rather to highlight some simple rules. First is the finding that the more compressed the growing period the higher the percentages of available labour used in farm work, accentuating the risk attaching to the more climatically marginal environments. However, while the proportionate commitment to crop management declines with increasing rainfall, intensification brings about a rapidly compensating increase in the amount of labour diverted into livestock management. More time spent during peak periods of demand for farm labour on feeding penned livestock is a necessary condition of intensification based on the use of animal manure.

Labour deficits which are especially frequent in the drier (more sharply peaked) systems translate into economic loss in the form of poor weeding, inadequate thinning, poor growth, unnecessary harvest losses, reduced yield and lower output per worker. Farmers are aware of such losses. However, the ambience of opportunistic farming does not dictate loss minimisation, but, on farms which normally include fallows, offers potential benefits from planting very large areas – at least when rainfall is good. Subsistence strategies include securing a grain output equivalent to two or three years' consumption where possible (Mortimore, 1989). Among the many uncertainties to be taken into account is the possibility of mobilising extra labour through hiring or exchanging arrangements. Thus a risk of 'strategic waste' is always present in the system, and this is quite often reflected in labour management.

Harvesting labour patterns show that with increasing rainfall, and a lengthening growing period, the complexity of harvest operations opens out from two to four peaks, each consisting of more than one major activity, separated from one another by transitions when the 'terms of labour' for competing tasks switch values. Negotiation of these must call for skilled management of the household labour force (and extra household labour, if available), and their anticipation in judicious choice of the planting dates for photoperiodic and non-photoperiodic crops. In an intensive system, the collection of fodder and residues may assume almost as important a place as that of harvesting the 'economic crops'.

Clearly, the idea of an 'average smallholder' throwing down a standard package of 'traditional' farm technology against the Sahelian elements is inappropriate, for neither does it happen in that way nor could it, given the variability and diversity we have outlined above. We have no hesitation in suggesting that smallholder farming under Sahelian conditions is more complex than most systems of large-scale commercial farming. Development interventions – projects or policies – have focused in the past on economic or technical constraints perceived to be facing African smallholder agriculture. Less attention has been paid to the intricacies of negotiating a way through the uncertainties of Sahelian rainy seasons, given capital and technological constraints. More attention may, with justification, be given to this angle.

5

WORKING NATURE

Resource endowments and change

In the short term, as our focus on flexibility and adaptability has underlined, smallholders in the Sahel are confronted by frequent changes in their resource endowments, some of them in variables over which they can exercise some control, but most of them not so. At any point in time, a man or woman is poised, as it were, before the shifting colours of a kaleidoscope, but only for a moment, before the picture changes. His or her decisions can affect the way it will change.

When this time perspective is extended into the past, the controversies about culpability for Sahelian degradation merge into the longer story of how communities, using their own labour, achieved the domestication of the Sahelian environments during the long span of human history in the region: a much more positive theme which clashes violently with contemporary evaluations of them as environmental spendthrifts, caught up in forces they cannot control. This contradiction is the subject of this chapter. In looking at some of the contemporary actions of humans as domesticators, we shall ask whether it is logical to assume that wisdom and skills that have developed over centuries (if not millennia) have no continuity in current or future practice, and deserve merely to be discarded.

Natural biodiversity

It is useful to put aside the idea of a staged evolution of culture from hunting and gathering, through pastoralism and to farming, and to recognise that almost all rural Sahelian communities practise all three, and have done so for a very long time. However, we shall not discuss wild faunal resources here, but restrict ourselves to the use of plant biodiversity in support of agropastoral households.

In Tumbau, all remnants of the natural Sudanian woodland, which was an open savanna dominated by thorny trees and perennial grasses, have now disappeared and been replaced by a mosaic of farmland (comprising over 85 per cent of the area) and shrubland, the latter occurring along cattle tracks and on degraded sites where ironstone crusts appear on the surface. In Dagaceri, drier, transitional woodland can be observed in nearby forest reserves, but within the village lands it too has been replaced by cultivated land (about 55 per cent of the area), grass and shrub fallows, and reserved grazing areas which contain a thin cover of annual grasses and widely distributed shrubs of *Boscia senegalensis, Guiera senegalensis* and *Hyphaene thebaica*. Futchimiram, by contrast, has extensive areas of mature Sahelian woodland, forming an open parkland of *Faidherbia albida* and *Balanites aegyptiaca*, under which annual grasses dominated by *Cenchrus biflorus* cover the entire surface. However, heavy grazing, continued for many years, is probably responsible for the observation that young trees tend to be scarce in an age-size structure heavily weighted in favour of older trees, casting doubt on the regenerative capacity of the woodland. Farmland is confined to about one-fifth of the area, and much of it shifts from time to time, leaving long fallows where (it must be said) woodland rapidly recovers. Kaska, as described in Chapter 3, lies in natural grassland (where *Cenchrus biflorus* is strongly dominant) in which woodland only occurs in the depressions. The grassland, though subject to reactivated dune formation, is less altered than the depressions, where most farmland (fluctuating between 10 and 15 per cent of the area) was concentrated until recently, to the detriment of *Hyphaene thebaica* woodland in particular. Thus the status of the natural vegetation is quite different in each of our locations, where the pressure on woodfuel and timber resources also varies sharply with the density of the population.

The natural habitats of West Africa have long provided a range of foods, fodders, medicines, craft and constructional materials. Their diversity attracted attention early in colonial times, provoking the authorities to sponsor a comprehensive inventory of West African useful plants (Dalziel, 1937), though official interest in their resource potential later waned. Plant biodiversity might not be expected to be high in the Sahel, in environments which suffer from both aridity and high levels of farming activity. However, a study in Dagaceri (Mohammed, 1994) recorded 121 non-domesticated plants which are used by the people under private, common, or open access (Table 5.1). Of these, 62 per cent are herbs, 28 per cent trees and 9 per cent shrubs. There are eight management regimes for these species, which are, in order of frequency, medicinal uses (32 per cent of species), cultural uses, fodder, food, construction, agriculture, fuelwood, shade and soil protection. The study concluded that both 'indigenous and induced management principles are all geared towards protection of the vegetation' (ibid.: 202).

Records of major famines in the Sahel show that when crops fail and food is scarce, the people increase their intake of gathered foods and extend the inventory of acceptable items to include a wide range of plants (Mortimore, 1989). During the Sahel drought of 1972–1974 in northern Kano, ten species in particular were used widely throughout a sample of 631 villages, such as the leaves of the tree *Adansonia digitata* and the forb *Cassia obtusifolia*. Another 37 species were identified and used in some places, and there are records in the

Table 5.1 Multiple use plants at Dagaceri, by land use class and access (no. of species)

Type of access	Settlements	Farmland	Fallows	Rangeland	Cattle tracks
Private	9	41	30	0	0
Common	1	0	23	0	0
Open	2	23	72	70	32

Source: Mohammed (1994: 125–128)

northern Nigerian literature of a further 21, whose use was unconfirmed at that time. The knowledge of these famine foods was transmitted to a considerable extent by the women, whose work it was to collect them. Whether collected from the bush or from the hedgerows and roadsides of the farmed landscape, some of these plants can supply vitamins and other substances needed to maintain health under conditions of malnutrition and exposure to the diseases of the wet season, a time when high work loads coincide with food shortages or the 'hungry season'. An extensive pharmacopoeia derived from indigenous plants in the Kano Close-Settled Zone has been documented and analysed (Etkin and Ross, 1982). This work has shown that the interrelations between 'medicine' and 'food' – categories that are not always easy to distinguish – have implications for the efficiency of the workforce during the farming season, as well as for health, and thereby the welfare of the community.

Crop biodiversity

All cultivars have been domesticated from wild relatives somewhere in the world. Relatives of pearl millet, sorghum and other less important crops are found in the grassland communities of the Sahel (Harlan, 1995). Combining these with domesticates introduced from elsewhere (for example, cotton, cassava or groundnuts), farmers have created an inventory of crops from which to make their adaptive selections for particular places and rainfall conditions. By frustrating natural ecological succession (through weeding their farms), farmers create a controlled ecosystem within which they manage their genetic resources – their crop plants – in order to optimise economic benefits.

During our fieldwork, inventories of the cultivars and landraces known and used as crops in each of the four villages were compiled. They were constructed in group discussions carried out during the harvest season, and supported by specimens brought in by farmers. As the object of the excercise was understood to be to achieve the most comprehensive inventory possible, the farmers in each place, as well as the interviewers, emphasised incremental information rather than analytical discussion of the types already identified.

The inventory produced 76 plants in Tumbau (both upland and lowland types), 48 in Dagaceri (upland only), 55 in Kaska (both upland and lowland), and 23 in Futchimiram (upland only) (Table 5.2). The inventories also included

Table 5.2 Cultivars in the four study villages

Cultivars	Number of landraces			
	Tumbau	Dagaceri	Kaska	Futchimiram
Cereals				
Benniseed (*Sesamum indicum*)	1	2	2	0
Maize (*Zea mayz*)	1	1	1	0
Pearl millet (*Pennisetum glaucum* subspecies *Americanum*, syn *P. typhoides*)	12	7	6	3
Rice (*Oryza sativa*)	1	0	0	0
Sorghum (*Sorghum bicolor*)	22	18	13	6
Wheat (*Triticum aestivum*)	(1)	0	1	0
Beans				
Cowpea (*Vigna unguiculata*)	9	6	9	7
Soya (*Glycine max*)	0	0	1	0
Earth nuts				
Bambara groundnut (*Voandzeia subterranea*)	3	4	1	1
Groundnut (*Arachis hypogaea*)	5	2	0	4
Roots				
Cassava (*Manihot esculentum*)	4	0	3	0
Sweet potato (*Ipomoea batatas*)	2	0	1	0
Melons				
'Cow melon' ('*guna*'; *Citrullus lanatus*)	0	2	4	0
Melon (*Citrullus vulgaris*)	0	0	1	0
Pumpkin	0	0	1	0
Water melon	0	1	1	0
Vegetables				
Karkashi	(1)	1	(1)	0
Okro or Okra (*Hibiscus esculentis*)	1	2	1	1
Onion (*Allium cepa*)	1	0	1	0
Peppers (*Capsicum* species)	5	0	1	0
Sorrel or Roselle (*Hibiscus sabdariffa*)	3	2	2	1
Tomato (*Lycopersicon esculentum*)	1	0	2	0
Sugar cane (*Saccharum officinarum*)	1	0	1	0
Industrial crops				
Cotton (*Gossypium hirsutum*)	0	0	1	0
Kenaf ('*rama*'; *Hibiscus cannabinus*)	2	0	0	0
Total landraces	76	48	55	23

Sources: Chiroma (1996); Ibrahim (1996); Mohammed (1996); Yusuf (1996); field books and integrated project inventories.

Note:
(1) No local record, but well known in the region.

farmers' perceptions of each one with regard to a number of standard characters. These included, in addition to its name(s) in the local languages, its origin (the area where it originated, or the person or extension agency responsible for its introduction), its popularity (as a crop to grow, as a food, and as a product for sale), its biological properties (growing time to maturity, morphological characteristics – colour, size, etc. – and resistance to pests and drought), its response to fertiliser and any restrictions on its cultivation imposed by soil.

Farmers recognise types at two levels: inter-specific (cultivars) and intraspecific (landraces). The first level presents no problems as the distinctions they make coincide with recognised taxonomic categories. However, the second level distinctions made by farmers do not necessarily derive from genetically significant differences, nor are they necessarily applied consistently from place to place or from year to year. The names (which of course use local languages) are assigned on the basis of observed morphological and phenological characters, and known origin (associations with places, persons or extension organisations). Any collection made in the field may reveal morphological diversity within a named landrace, especially where cereal grains are concerned. Deciding equivalences between similar landraces having different names or differing landraces having similar names is only possible in an approximate sense, given the diversity both in observed (morphological or phenological) characters and in their genetic base (discussed below). The concept of 'inventory' with which we started out – relying on indigenous taxonomies and terms – therefore has a rather limited value. We use the word 'landrace' to mean a population of plants having the same name in the local language and distinguished from populations having other names (Bunting and Pickersgill, 1996).

A preliminary review of these data indicates that there is a positive correlation between ecology and diversity. Tumbau is the richest, and Futchimiram the poorest, in terms of the domesticated biodiversity available. However, this is not merely a function of rainfall. Tumbau and Kaska have lowland soils where dry season cultivation, with or without irrigation, is possible. Wheat (*not* rice), several vegetables, some melons and sugar cane depend on such soils in Tumbau. Kaska is able to compensate for low rainfall with its lowland soils, where maize, wheat, soya, some melons, roots, peppers, some vegetables, sugar and cotton can be grown in small quantities. There is also a positive correlation between rainfall and the numbers of millet and sorghum landraces available, but significantly not with numbers of rainfed cowpeas and groundnuts, which are comparable in Tumbau and the drier villages.

The numbers of cultivars and landraces being maintained in all these systems is testimony to the advantages of diversity, for smallholders with a subsistence priority, and the weak development of market-based specialisation, using comparative advantage. However, the number of crops and landraces gives no indication of their relative importance in production, either for domestic use or for the market. Qualitative indicators of their popularity suggest that a few provide the greater part of output. For those cultivars (millet, sorghum, cowpea) having the largest numbers of landraces available, this raises the question: How and why do farmers maintain seed stocks from year to year for landraces that are rarely used? The fact that these stocks are maintained suggests that a high value is placed on domesticated biodiversity as a resource. This supposition is supported by the variable adaptation of landraces to rainfall and soil moisture conditions, themselves highly variable in space and time.

An additional dimension to understanding diversity in crop inventories is the position of farming in the household economy. In Futchimiram, livestock including cattle play a larger role in household economies, and unlike Tumbau, where small ruminants may be said to support the cropping system through a high degree of integration, in this system cropping supports livestock by supplying residues at a lower level of integration. The lower level of cultivar diversity in Futchimiram reflects, to an extent that is not easy to assess, a different pattern of economic specialisation.

Diversity in pearl millet (*Pennisetum glaucum* subspecies *americanum*, syn *P. typhoides*)

Pearl millet is the staple grain in all four villages, though in Tumbau its position is contested by sorghum; in Futchimiram very little sorghum is grown. Work on the genetic mapping of pearl millet is well advanced. Millets are known to have a high phenotypic diversity (Wilson *et al.*, 1990), outcrossing freely with related wild grasses. It is therefore important to know to what extent farmers in the Sahel are able to manage this diversity effectively in order to optimise crop resistance to problems such as variable rainy seasons or pest attack, and to what extent they do so as part of a self-conscious strategy. This question was addressed through linked studies of the genetics of certain millet landraces (Busso *et al.*, forthcoming), and an analysis of the understanding and choice of crop plant characteristics and the management of seeds.

A landrace is not a taxonomic entity but a dynamic artefact (Bunting, 1992). Cultivated pearl millet outcrosses readily with the wild *P. glaucum* subspecies *monodii* (Brunken, 1977). In practical terms this means that domestication is not something achieved several millennia ago so much as a continuing struggle, fought year by year on each smallholding. Seed management (which includes selection, storage, planting and weeding practice) is therefore critical to the manipulation of the genetic resources. With regard to the first of these, Harlan (1995: 34) has generalised:

> All of the selection pressures push domesticates in the same direction: reduced dormancy, reduced appendages, larger seed, more synchronous ripening, nonshattering, recovery of fertility in reduced or rudimentary flowers, greater fitness for the environments provided, including adaptation to climate, soils, pests and diseases, cultural practices and selection by the growers.

80

Plate 5.1 Managing genetic resources: local landraces of pearl millet, Dagaceri: *lafsir* and *idon hawainya*.

Small farmers in the Sahel select millet heads in the field according to landrace (which has known performance and product characteristics), and individual variables which include spike length and thickness, and grain colour and size. The effect of selection is to reduce genetic diversity within, and to increase diversity between, landraces; and the effects of on-farm outcrossing are the opposite. Most farmers either select from their own fields or obtain seed from trusted neighbours. The exceptions to this rule are when a new landrace is introduced to the locality, and when seed is scarce (for example, in drought) and must be bought from food grain sellers in the market (which implies an unknown provenance). Under normal conditions, therefore, there is a strong localising tendency at the level of both the village and the farm.

The choice of landraces by farmers is also dynamic, as they attempt to adjust to rainfall from year to year, soil moisture and nutrients from site to site, interactions with other crops in the mixture, market prices, resistance to pests and diseases, labour supply, and taste, colour or cooking properties. In the two villages used for this detailed crop diversity study, one late maturing and seven early pearl millet landraces are identified. However, at least two of these are hybrids from outcrossing. The seed stocks of the selected landraces are reproduced within the locality from year to year.

Analyses of the genetic diversity in the cultivated millet landraces were concentrated in the village of Dagaceri, with supplementary sampling in Kaska (90 km NE of Dagaceri) to test for inter-village effects. Three pearl millet landraces were selected:

- *badenji*, a standard early type originating in the region; seed selected from each year's crop;
- *dan arba'in*, a very early type introduced within the last 20 years; seed selected;
- *lafsir*, a hairy local wild or hybrid type, not selected, but appearing randomly in the crop.

The sample design stratified the material according to village, farmer's field and landrace. In Dagaceri, the samples were collected in the field from one stand in ten, and from one plant (spike) per stand. Three landraces were collected from three farmers' fields. Four spikes per landrace were selected for analysis and six replications (grains) from each spike. In Kaska, one landrace per farmer's field was provided by the farmer. Using an amplified fragment length polymorphism (AFLP) technique, 288 plant samples were analysed for 252 molecular markers. Comparisons of diversity were made using principal components analysis and analysis of variance (Busso *et al.*, forthcoming).

The principal components were subjected to analysis of variance, quantifying the factors used in the experimental design: region (village), farmer (field), landrace, head (spike) and seed. The statistically significant associations are set out below.

Landrace

The factor 'landrace' is highly significant. *Lafsir* is the most diverse and *badenji* the least. *Lafsir* is a hybrid and is not selected. *Dan arba'in*, however, is no less diverse internally than *badenji*, even though it originated from research and extension and is the most recently introduced to the area (about 20 years ago). Since then it may be presumed to have become more diverse through outcrossing.

Farmer

This is the most significant factor identified. When each landrace is plotted by a farmer's field *in the same village*, the variability between farmers is greater than the variability between landraces. This indicates a degree of independence between farmers' gene pools which is consistent with their practice of deriving their seed each year either from their own field or from a regular source. Farmers have a complex understanding of the ecological niches in their locality, accumulated over many years, as noted by Gadgil *et al.* (1993). Selection pressure exerted at the level of the individual farmer, as well as that of the region, may be expected to amplify the diversity of cultivated millets noted by Brunken (1977). 'Farming *is* plant breeding' (Harlan, 1995: 36).

Heads, seeds

Variability was also found between the sampled heads. It was predictable on the basis of the outcrossing capability of pearl millet, and the presence in the environment of wild and hybrid relatives.

The unexpected conclusions of this work are that the variability of the pearl millet populations is affected, first, by farmers' seed management (creating partly independent gene pools on each farm); second, by regional isolation (as seed is only transported any distance from its source under certain circumstances), and third, by landrace (which, on the basis of phenotypic characters, was expected to be the most important). This study was based on only one year's data, so variability between years could not be investigated. The level of variability is similar among heads, landraces and farmers.

It would be a mistake to make too sweeping a generalisation about farmers' ability to act as their own 'crop breeders', creating homogeneous landraces whose characteristics are stable between one household or village and the next, or one year and the next. The demonstrated diversity is greater than this simplistic idea would imply. However, each farmer does *manage* the genetic resources at his or her disposal. Notwithstanding the known importance of outcrossing, both between adjacent fields and between domesticated landraces and wild relatives, farmers can control the genetic flux sufficiently to build and maintain a set of desired characters for their crops. They do so without constraining the genetic diversity of the system very tightly. It is a nice question as to whether the constraints imposed on diversity by formal crop breeding (and even more so by genetic engineering), in the pursuit of greater productivity, offer a more sustainable pathway for Sahelian farmers than their present looser regime of genetic management.

Fodder resources

The natural grasslands have been grazed, initially by wild animals and later by domestic herds and flocks, for so long that the effects of grazing must be considered as an integral part of the savanna ecosystems. In Futchimiram, where there is the most woodland and grassland (covering three-quarters of the area), heavy grazing tends to eliminate tree seedlings, affecting regeneration, and creates the distinctive 'parkland profile' wherein foliage is removed from the

Plate 5.2 Managing fodder resources: Badowoi cattle in *Cenchrus biflorus* annual grass-land near Futchimiram, July 1992.

lower canopy, giving a horizon in the landscape at the height which the largest animals can reach. Where branches are cut down to supply browse in time of fodder shortage (a practice which is associated with nomadic herders), the canopy area is adversely affected, and it is even known for trees to die.

If these effects of grazing on trees are recognised, its impact on the grass and herb communities is less clear. Owing to the resilience of Sahelian species, it is difficult to be sure whether eating grasses before seed formation materially affects their regenerative capacity. Seed has been known to survive for years before germination under arid conditions. Similarly the impact of selective grazing is hard to assess. An a priori case can be made that preferences, on the part of free grazing livestock, for certain palatable species, if exerted consistently, can increase the competitive advantage of less palatable ones, thereby reducing the stock-carrying capacity of the pasture in any one year. However, graziers in our villages say that the patterns of specific dominance which appear every year in particular places are variable, and unpredictable on the basis of previous years.

Superimposed on these management effects is the impact of rainfall change. A transition from perennial grasses to annuals has been observed in the arid Sahel, including on ranches where grazing is strictly controlled. In the Manga Grasslands where Kaska is situated, the dominant grass species before the Sahel drought were perennials, including *Andropogon gayanus* (gamba grass), according to resource surveys (LRD, 1972). Since 1977, if not earlier, annuals (principally *Cenchrus biflorus*) have been dominant. Such a transition appears unlikely to be due to selective preferences, as the annuals are actually more palatable during the late dry season when the resident livestock rely on the grassland, than the thick, woody stems of the perennials. Technical assessment of the stock-carrying capacity of the two communities indicates that the annuals are in no way inferior (Rippstein *et al.*, 1972), a paradoxical finding in relation to the once common view that the transition was a form of ecological degradation.

This brings us back to rainfall variability as a fundamental determinant of economic productivity in these environments, making questions of stock-carrying capacity extremely complex (de Leeuw, 1996). The new paradigm of range management by African herders in arid areas advances our understanding from old ideas based on concepts of equilibrium to a more adaptive model of opportunistic stocking, in which the vagaries of rainfall (and of natural biomass) are followed by disequilibrial stocking strategies, multiplying animals to take advantage of fodder resources in good years, and suffering losses in the bad (Sandford, 1983; Ellis and Swift, 1988; Behnke *et al.*, 1993).

One strategy for responding to periodic or local fodder scarcity is to move herds and flocks around. Nomadic FulBe herders pass through all our villages from time to time, but they are few in number compared with the sedentary or semi-sedentary FulBe. Some of these also send their cattle south on transhumance from the densely settled farmlands of Tumbau and the congested grazing reserves of Dagaceri. Transhumance in northern Nigeria is doomed in the longer term as more grazing lands fill up with farmers, and more wetlands are appropriated for irrigation. But in Kaska and Futchimiram, transhumance has never been of major importance for the resident FulBe and Badowoi of those areas. They rely on the annual grasslands, and suffer their sometimes huge losses when times are hard. Questioning has elicited a surprising ignorance of alternative grazing resources outside the area.

Emergencies may, however, disturb such patterns: for strategic behaviour is not adaptive if it cannot devise new solutions. In the drought year of 1984, the grazing lands around Dagaceri were invaded by nomadic FulBe from near Gure in Niger. They had suffered levels of mortality unknown even in the Sahel Drought, and decided on an unprecedented migration into Nigeria, a territory previously unknown to them. As ethnic relations of the resident FulBe, they were not obstructed from access to the pastures, which their large herds consumed over several weeks of grazing before moving on. The knock-on effect was, of course, an undermining of the local stock keepers' resources. The clash of priorities between claimants to fodder resources which can occur as a result of herd movements is less amicably illustrated in outbreaks of conflict which frequently erupt along the Hadejia River in north-east Nigeria, and in the destruction inflicted on private trees standing on farmland in the Kano Close-Settled Zone, a form of access which local Hausa farmers feel themselves powerless to stop.

With the expansion of farming, more and more crop residues became available for animals, and as their value has increased, free access has increasingly been denied. Farmers in Tumbau hoist bundles of residues into the forks of trees to prevent access by free-grazing animals in the dry season and to be used later for feeding their own livestock. Contracts exchanging manure for access to residues are still agreed in some areas, but it is increasingly common for money to change hands as well. Farmers in Dagaceri, when they anticipate a fodder shortage, go out into their fallows and cut and carry the grass for storage in bundles (*budu*) and later sale – a practice condemned by the FulBe livestock breeders who regard other people's fallows as within their access rights. An increasing scarcity of natural grazings relative to farmland sources of fodder shifts the balance of advantage in livestock keeping towards farmers, whatever the trends may be in the local terms of trade for crop versus livestock products. In Tumbau we can observe one end of a continuum, where natural grazings have been all but eliminated, and farmers succeed in maintaining livestock (mostly, but not all, small ruminants) at densities higher than anywhere else in northern Nigeria. Plant biomass productivity averaged 7 tons/ha on three farms in 1994, of which 1.7 tons/ha represented crop output, 2.4 tons/ha were used as fodder, and the rest supplied construction materials or 'waste' (Harris and Bache, 1995: Figure 11).

In farming systems which combine livestock with cropping, the management of fodder resources, whether natural grazing or farm residues, adds to the complexity introduced by the development of cultivars. In some intensive systems in Africa, fodder crops are grown, for example, on the terrace banks of Machakos in Kenya (Tiffen *et al.*, 1994). In our systems in Nigeria, it seems doubtful if land can be devoted exclusively to fodder production when so much edible biomass is generated as a byproduct of the cropping system.

There remains a third dimension to the story of the domestication of the natural environment.

Trees

Trees are managed (or mismanaged) not only in natural woodland but also on farmland. The neglect of this dimension of silviculture has led to many unforgivable condemnations of African attitudes to trees and their conservation on the part of government agencies and international observers (Cline-Cole *et al.*, 1990; Cline-Cole, 1997). Tree management is embedded in a complex political economy involving the conflicting interests of centralised agencies (set up by colonial governments to enforce forest conservation), woodcutting entrepreneurs, self-aggrandising local governments, and the remnants of local autonomy over resource management. These struggles hide behind local realities which can be observed in terms of objective indicators such as tree densities or species distributions. The fundamental process which is at work, however, is the transformation of natural woodland into a new land use mosaic in which trees,

Plate 5.3 Managing trees on the farm: *Ceiba pentandra* over pearl millet, *Faidherbia albida* (leafless during the rains), and *Parkia biglobosa* (distant), at Tumbau.

contrary to the claims of those who see farming and deforestation as natural bedfellows, are being managed in new ways. A review of the place of trees in each of our study areas now follows.

Tumbau

The elimination of the natural vegetation in the northern savanna, and its replacement by a cultural vegetation, or 'farmed parkland' (Pullan, 1974), has gone furthest here. Effectively, the farmers have domesticated all the trees and plants in the system, not just their crops. There are over 40 different multi-purpose farm trees found in the area, most of them having from two to four uses, and some shrubs. All are used for fuelwood in one way or another; many provide fodder; a number provide fruits; and some are cut for timber. Trees are most numerous (per hectare) near the villages and within them (shade trees), while shrubs are more commonly found on farmland further away. Fuelwood is mostly harvested from living trees and not obtained by felling, though selling a tree to a woodcutter for felling is a recognised if infrequent strategy. Trees contribute to feeding livestock and thereby to manure production for farming; the end-product of fuelwood is ash, which is subsequently taken to the farms as manure.

There is not much planting or transplanting of farm trees (though the techniques are well known), partly because few trees are felled. The greatest effort in planting and raising trees is for shade and timber species in the villages. Most farm trees regenerate spontaneously, and the farmers leave them and nurse them to maturity. Some seedlings (for example, *Parkia biglobosa, Mangifera indica, Ceiba pentandra* and date palms) need to be fenced off in the dry season to protect them from the free-grazing livestock. Other trees grow from shrubs without assistance. The fact that fuelwood may not be taken from other people's economic trees protects them from unscrupulous lopping.

The average densities of mature trees on farmlands in the Close-Settled Zone have been estimated (from air photographs and small ground surveys) to be 12–15/ha in the west and 7–9/ha in the east (Mortimore *et al.*, 1990; Nichol, 1990). In terms of timber volume the two areas are much the same, as the density is influenced by species and the size of individual trees. The largest species is *Adansonia digitata*, which can dwarf all other trees in its neighbourhood. Studies of tree densities over a period of time which included the Sahel drought show stable or increasing densities, notwithstanding the moisture stress that caused a number of individuals to die, and the pressure on hungry households to cut their trees and sell them as fuelwood in burgeoning urban markets.

Dagaceri

Of the non-domesticated species inventoried in Dagaceri (see Table 5.1 on p. 77), 28 per cent are trees and 9 per cent are shrubs. Villagers are aware of the significance of vegetation and they have taken various strategic measures for its management. The local government has made free distributions of the exotic neem (*Azadirachta indica*) seedlings and other introduced and indigenous species. Fuelwood comes from dead wood, dry, woody shrubs such as *Guiera senegalensis, Boscia senegalensis* and *Acacia ataxacantha,* and crop residues such as millet stalks and sorghum. Living trees are rarely felled. Various economic products are harvested from trees, such as the leaves of the baobab, *Adansonia digitata*, which is the most prominent tree on farmlands (harvested in both the rainy and dry season), the leaves of *Faidherbia albida* (cut in the dry season for livestock), and of *Combretum glutinosum* (cut for livestock in the dry season). Fruits of *Hyphaene thebaica, Ziziphus spp., Balanites aegyptiaca* and *Tamarindus indica* are harvested in the dry season.

While the value of trees and the need to conserve them on farmland is recognised in Dagaceri as in Tumbau, the distribution and density of trees betrays an immature stage in the evolution of farmed parkland. On the permanent fields close to the village, densities attain 3–5/ha, a much lower order than in the Kano Close-Settled Zone. However, the lower density takes no account of the fact that many of the trees are large individuals of *Adansonia digitata* whose visual and ecological impact is greater than their numbers suggest. Further out of the village, on land which is subject to fallow cycles, shrubs rather than trees dominate. On grazing land, shrubs also dominate and mature trees are widely spaced, except for rare thickets in wetter depressions. Compared with the woodland which may be presumed to have grown before farmers arrived in the area, and is visible in nearby forest reserves, this is indeed a deforested landscape. But an imminent transition is suggested by the trees on farmland, and by observations of increasing numbers of trees during the past two decades.

Futchimiram

The natural woodland still contains a copious supply of trees, especially *Faidherbia albida* and *Balanites aegyptiaca*. The spatial distribution of trees, which is uneven, is affected by cultivation, fallow, fire, grazing by animals and by drought, which has become a persistent feature in the area. The Badowoi do not plant trees to replenish those destroyed. They have no need. The dispersed populations seen in the fields are not selected and planted species, but are protected volunteers. However, when farmers clear new fields, they do not fell the existing trees but instead pollard them and use the branches for fencing. They believe that canopied trees attract bird pests to farmland by providing perches from which they can demolish the emerging millet. After farmland has returned to fallow, the trees recover their canopies. Hence the density of trunks seen on farmland and in woodland may be quite comparable, though the canopy cover diverges sharply. Shade trees and the exotic, *Azadirachta indica*, though popular elsewhere, are less common in Badowoi villages.

Tree felling has been reduced since around 1990. Forest regulations, introduced under the influence of anti-desertification programmes, state that trees on farmland may be cut, but in the bush *(Kanuri: kara'a)* only branches may be cut, and then not to the extent of endangering the life of the tree. The law applies to all trees, however small they may be. A majority of farmers claim that the law is effective and is leading to an increase in woody vegetation. Women, often assisted by children, collect fuelwood, particularly dead twigs and branches. They may fell dead trees by setting them alight. There is no shortage of fuelwood.

Kaska

The situation of woodland is entirely different in Kaska owing to its localisation in the depressions. There are three stories to tell.

First, the mixed woodland (*Acacia sieberiana, Piliostigma thoningii*, with individuals of *Hyphaene thebaica*) which appears occasionally in depressions and is related to the woodland found in the ecozones surrounding the Manga Grasslands, takes the form of dense thickets which may have lost out to *Hyphaene thebaica* woodland under the influence of increasing salinity. If so, it is a residual type used for fuelwood cutting.

Second, the *Hyphaene thebaica* woodlands, which surround the *tafkis* in dense, sometimes almost impenetrable stands excluding all other forms of land use, and also appear dispersed in the *faya* depressions, have been extensively felled, as is shown by the better survival of these woodlands in Niger where forestry law was more strictly enforced. Although the fruit has little value (though edible), the fans of this tree supply fuel requirements, the trunks are the main construction material for roof supports, fences and well-heads, and the regenerating plants produce fibrous shoots (*kaba*) which are harvested (by men) for sale and for manufacturing mats and ropes. Intensive harvesting must certainly inhibit regeneration, as the trees cannot grow to maturity which, with continued felling, threatens this woodland, which is under open access.

Third, unlike in the other villages, economic trees are not an integral part of rainfed farms. Soil moisture conditions are marginal, especially on the *tudu*, where several government shelter belts planted against desertification in the 1980s have come to nothing. Rare exceptions are provided by solitary individuals of *Faidherbia albida*. However, in the *kwari* depressions among the irrigated *garka* farms, and near villages, there are small date palm plantations and some citrus trees. Date farming is a skilled speciality in the area, the trees are privately owned, and income from it adds significantly to the sustainability of the farming system.

Fuel in Kaska comes from trees in the lowland depressions and around settlements (*Azadirachta indica, Balanites aegyptiaca*). *Hyphaene thebaica* and date palms provide poor fuel, but due to measures taken by the local government authority against cutting live trees, the stalks and fans can be used for domestic burning. These are usually collected by women.

Landscape transformation

Armed with these resources of natural, managed and cultivated plants, the four communities have added their labour to create, in countless small increments, a new landscape reflecting an optimal spatial allocation of land. Using air photographs, ten major types can be distinguished, which are described as follows:

Upland

1 *Arable land* is rainfed farmland with some protected trees (of economic value) at variable densities. It includes (in Dagaceri and Futchimiram) fields where weeding was abandoned, and short grass fallows.
2 *Dense woodland* is a natural community dominated by trees or shrubs with a closed (or nearly closed) canopy.
3 *Open woodland and grassland* includes a range from mature woodland with a broken canopy and annual grasses (Futchimiram), through naturally regenerating shrubland with annual grasses (Dagaceri), and open grassland of annual grasses and forbs with no trees (Kaska), to a thorny scrub with

thin perennial grasses (Tumbau). The variety of communities makes finer distinctions impracticable.

4 *Sparse vegetation* is bare ground (as it appears on air photographs, which are taken in the dry season), with a few shrubs. When occurring on rock, laterite or hardpan, it suggests degradation; however, on air photographs, cultivated land sometimes offers a similar reflectance.

5 *Active dunes* only occur in Kaska.

6 *Settlements* exclude the dispersed compounds associated in particular with FulBe herders. They support woodland, with shade trees along the streets and, inside the compounds, some indigenous fruit and other trees.

Lowland

7 *Arable land* is cropland in wetlands or depressions, where additional soil moisture, or irrigation, permits perennial or irrigated crops.

8 *Dense woodland* is found mainly in Kaska, where closed canopy forest of *Hyphaene thebaica* surrounds the lake beds in the deeper depressions.

9 *Grassland* is seasonally flooded grassland, usually without trees, in depressions.

10 *Water or wetland* is only associated with rivers in Tumbau; elsewhere it is seasonal ponds or dry beds with (usually) an unvegetated rim.

Upland (or *tudu*) comprises the greater part of all four study areas (see Chapter 3). *Lowland* consists of the flood plain of the Kano River and small seasonal tributaries (Tumbau), a few small depressions with seasonal ponds (Dagaceri and Futchimiram), and deep depressions occupied by saline ponds or lakes (*kwari* depressions, Kaska). The *faya* depressions of Kaska are included under *upland* as they do not have shallow water-tables (though the distinction is sometimes hard to make).

A mosaic of these types is increasingly dominated (in most areas) by arable (or cultivated) land, which diminishes along the transect north-eastwards, from over 85 per cent of the area in Tumbau to less than 15 per cent in Kaska. As the transformation of land from natural vegetation to cultivation is achieved through labour, with little or no assistance from machines, this percentage is a critical indicator of the intensity of the relations between society and environment.

Mapping from vertical panchromatic air photographs was carried out in four areas, each of approximately 100 km^2, and at three points in time (Table 5.3). However, such sequential air photo interpretation is subject to a degree of uncertainty which arises from differences in scale, resolution, distortion and tone among differing sets of photographs. The problem is compounded when sites of differing ecology are subjected to a harmonised classification. Furthermore, the interpretation is influenced by the effects of the rainfall during the season preceding the photography, which was taken during the period September

Table 5.3 Land use types in four areas (%)

Mean rainfall, 1992–1996 (mm), farming season	Tumbau 533			Dagaceri 360			Futchimiram 326			Kaska 301		
Year	1950	1971	1981	1950	1969	1981	1957	1969	1990	1950	1969	1990
(a)	810	698	573	706	384	429	461	224	286	557	224	286
(1)	75.8	87.2	87.7	35.6	56.1	54.6	22.1	22.5	21.7	0.9	0.4	10.0
(2)	0.2	0.3	0.6	3.2	1.6	1.0	0	0	0	0	0	0
(3)	12.6	4.1	0.3	57.1	35.4	32.7	77.0	77.1	70.0	73.0	87.2	56.6
(4)	4.1	3.9	7.5	1.9	5.2	11.1	0.6	0.2	7.5	0.6	0.9	5.1
(5)	0	0	0	0	0	0	0.9	0.9	20.2	0	0	0
(6)	0.3	1.6	2.8	0.3	0.4	0.5	0.2	0.2	0.8	0.2	0.1	0.2
(7)	1.8	2.1	0.7	0	0	0	0	0	0	14.9	3.8	0.7
(8)	0	0	0	0	0	0	0	0	0	5.4	4.1	4.6
(9)	4.8	0.5	0.3	0	0	0	0	0	0	1.6	0.8	1.0
(10)	0.4	0.3	0.2	1.9	1.3	0.1	0	0	0	2.5	1.9	1.6
(11)	77.6	89.3	88.4	57.1	35.4	32.7	22.1	22.5	21.7	15.8	4.2	10.7

Notes

Explanation of Column 1

(a) Annual rainfall recorded at nearest synoptic station in the year preceding photography: Kano (for Tumbau), Nguru (for Dagaceri), Maine Soroa (for Futchimiram and Kaska).

Land use classes:

(1) Arable land (upland); (2) Dense woodland (upland); (3) Open woodland and grassland (upland); (4) Sparse vegetation (upland or lowland); (5) Active dunes (upland); (6) Settlements (upland); (7) Arable land (lowland); (8) Dense woodland (lowland); (9) Grassland (lowland); (10) Water or wetland (lowland); (11) All arable (upland plus lowland)

to November. The materials available for the four sites end in 1981 at Tumbau and Dagaceri, and in 1990 at Futchimiram and Kaska.

It is clear from Table 5.3 that land use change has followed a different course in each area (see Chapter 3 for descriptions of the systems).

- *In Tumbau,* the percentage of arable (both upland and lowland) remained high throughout the period from 1950 to 1981 (and has continued so since), though evidently there was a small residue of uncultivated land that was appropriated during the first few years. Expansion thereafter was constrained by the simple fact that there was no more cultivable land available. In fact a small decrease occurred after 1971– entirely in the lowland, where drought, and the construction of dams upstream, inhibited irrigation in the flood plain of the Kano River.

- *In Dagaceri,* by contrast, Figure 5.1 shows that the area of arable land rose rapidly between 1950 and 1969, but then stabilised, even though more than 40 per cent remained uncultivated. This was mainly due to administrative measures to protect rangeland from further encroachment, as the land use boundary reflected an ethnic divide between Manga and FulBe. There was a decrease in the amount of dense woodland and an increase in the sparsely vegetated areas – both indicators of increasing pressure on the land. Since 1981, after the latest available photography, more fields have passed from short fallowing to annual cultivation, and trees have increased in number on farmland near the village.

- *In Futchimiram,* the area of arable land remained remarkably constant – at around 22 per cent – from 1957 to 1990, indicating little long-term change in the mix of annual cultivation and long fallowing which characterises this system. A late increase in sparsely vegetated areas, and evidence of diminishing tree densities in some places (not shown in the table) suggest, however, that all may not be well.

- *In Kaska,* with its more complex topography, a fluctuating arable percentage can be put down, in part, to shifting cultivation (and in part to interpretation difficulties). Desiccation after 1969 (the Sahel Drought) made some lowland soils difficult to manage and cultivation expanded on to the upland, where a rapid increase in moving dunes has since eaten away at the land available either for cultivation or for grazing.

Many of the changes recorded in this analysis reflect transfers from less to more intensively managed land uses. Work in progress suggests that cultivated land may be as productive as the secondary vegetation communities it replaces. But capitalisation (for such it is) usually involves some wastage of low productive resources, and these systems are not exceptions. Some degradation is suggested not only in the spectacular – if localised – increase in the area of moving dunes

93

DAGACERI

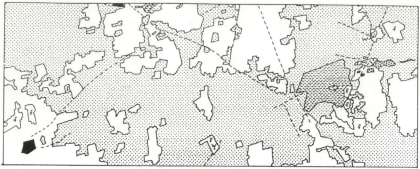

1950

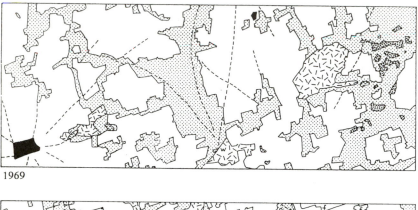

1969

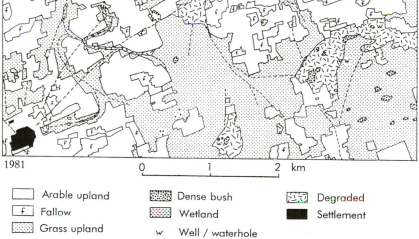

1981

0   1   2   km

	Arable upland		Dense bush		Degraded
f	Fallow		Wetland		Settlement
	Grass upland	w	Well / waterhole		

Figure 5.1 Land use change in Dagaceri, 1950, 1969, 1981.

at Kaska, but in increasing sparsely vegetated land, diminishing dense woodland, and even smaller areas of wetland.

However, a glance at the rainfall recorded in the years of photography (Table 5.3) provides evidence fully consistent with the thesis that the major factor contributing to degradation is not land use change, but rainfall diminution. The year 1950 was very wet (115 mm above the mean for 1961–1990 at Kano), and all subsequent years of photography were significantly drier. Such a thesis conforms with the judgements of farmers and stockraisers (Mortimore, 1989).

Conclusion

In this chapter, we have reviewed the ways in which labour and skills (social resources: Chapter 2) have been employed to domesticate natural landscapes, and the outcomes of these efforts. The creation of the 'farmed parkland' of the Kano Close-Settled Zone offers the most conspicuous example, perhaps analogous to that described for the forest–savanna transition zone of Guinea by Fairhead and Leach (1996). However, even the drier landscapes are also manipulated ecosystems, which have been enriched by farmers and livestock keepers, steered through the various endogenous and exogenous processes of change by human action.

Our research has taken us into inventorying the natural and human-made biodiversity in our four villages. We have restricted ourselves to plant biodiversity; another story remains to be told of the natural fauna and domestic animals. The place of crops is critically important in the productivity of Sahelian farming systems and, in order to understand the management of this resource better, we have introduced techniques of genetic analysis to pearl millet, the staple food crop of the region and an ancient, indigenous cultivar. Linking the indigenous knowledge and management skills of our farmers with molecular genetics has produced some new insights into the nature and impact of the process of domestication.

Crop resources, however, are not the farmers' only weapon; we have thus reviewed their management of fodder resources, and their transformation of natural woodland into farmed parkland. The sum of these activities is expressed in the changing map of vegetation and land use types which, traced over time and compared between villages, provides in summary form a diversity of resolutions to the challenges facing Sahelian communities in the management of their natural resources. These resolutions are investments in biological and technical knowledge, skills and labour time, which have been made for the most part in small increments, and by large numbers of individuals and households. In this they diverge sharply from both massive public sector investments (such as dams and irrigation schemes, often thought to hold the key to agricultural development) and private industrial or commercial projects. However, they share one thing with many other forms of investment: a small, unproductive fraction of the resource is discarded and left to degrade.

6

MAKING THE LAND
WORK HARDER

Towards intensification

It may appear paradoxical that, in the Sahel of all places, we should now embark on an analysis of how smallholders can be holding degradation at bay and improve the productivity of their dry, sandy and easily eroded soils. But the paradox is rooted in mistaken conceptions that the Sahel has overshot the limits imposed on its productivity by natural constraints and that smallholders are destructive consumers of biological resources – necessarily so, owing either to an ignorance of better methods or to an inability to change those methods. This is the established narrative (Swift, 1996) or 'myth' (Thomas and Middleton, 1994) of desertification, and its power to mislead the unwary (particularly unwary but powerful policy makers) is now recognised (Warren, 1996). If there is to be a future for the rural Sahel, these conceptions must be challenged, and with them, the apparent paradox of agropastoralists actually improving the productivity of their resources.

Agricultural systems with high densities of population have, indeed, achieved higher productivity per hectare in several parts of Africa and, in some if not all, sustained such higher levels for periods of decades or much longer (Turner *et al.*, 1993). Increasing population densities, translated into a scarcity of productive land, force the pace of agricultural intensification whereby cultivators (if not pastoralists) increase labour inputs per hectare to or beyond the point of diminishing returns, and exploit known technological options (whether 'indigenous' or 'exogenous' in origin) in order to maintain an upward pathway. Such technical change (which includes maintaining or enhancing soil fertility) is a condition for sustained development of output and incomes, 'agricultural growth' according to Boserup (1965). In investigating the reasons for the apparent success of such systems in maintaining productivity per hectare, recent studies have emphasised the use of labour to create conservation structures, integrate crop with livestock husbandry, improve soil fertility, develop on-farm tree husbandry and small-scale irrigation works, and raise the value of output per hectare (Cline-Cole *et al.*, 1990; Mortimore and Turner, 1993; Adams,

1992; McIntire *et al.*, 1992; Mortimore, 1993a; Turner *et al.*, 1993; Cleaver and Schreiber, 1994; Tiffen *et al.*, 1994; Reij *et al.*, 1996).

Tiffen *et al.* (1994: 29) define intensification as follows:

> increased average inputs of labour or capital on a smallholding, either on cultivated land alone, or on cultivated and grazing land, for the purpose of increasing the value of output per hectare.

This definition can be amplified to include:

- increases in labour inputs per hectare;
- creation of 'landesque capital' (Brookfield, 1984), for example, in the form of soil and water conservation structures, irrigation systems or enclosures;
- changes in technologies and in the crops selected;
- increases in the percentage of land cultivated and the frequency of cultivation cycles;
- more intensive management of trees and rangeland;
- more intensive methods of managing soil fertility, and integration of crop with livestock husbandry.

During the 1970s and 1980s it was expected among development agencies that African smallholders would follow the example of other continents into an increased dependency on inorganic fertilisers as the primary means of intensification. Countless agricultural development programmes (such as those funded by the World Bank in northern Nigeria: Wallace, 1980), policy recommendations to international development agencies (e.g. Breman, 1990), nutrient management models of mixed farming systems (Powell *et al.*, 1995), and 'balance sheets' of nutrients under 'soil mining' strategies of crop production (Smaling, 1993; van der Pol, 1992) – not to mention the obvious interests of international fertiliser companies – have converged on a powerful case for a form of output enhancement which has worked well for western, commercial systems.

However, even at the height of agricultural and 'integrated rural development' interventions, the adoption of this path was restricted to a minority of crops and farmers. The majority of farmers continued to rely unassisted on an 'indigenous intensification' pathway (Adams and Mortimore, 1997) which places major reliance on labour and low cost technologies, many of which (as Boserup hypothesised) are already known. Since the 1980s, devalued currencies, rising input costs, and stagnating product prices have caused capital-based intensification to falter in many countries, and underlined the strategic importance of 'indigenous intensification' potentials.

For many years, it was assumed that indigenous technologies were candidates for urgent transformation, as implied in the title of Schultz's influential work,

Transforming traditional agriculture (1964). It seemed unnecessary for researchers to spend time in analysing the ways whereby smallholders sought to ameliorate fertility or increase productivity under conditions of capital scarcity, even though such capabilities were known in the formal western literature (e.g. Stamp, 1938). Although the potential of indigenous technologies to adapt to change is now widely recognised in West Africa (Richards, 1985, 1986; Reij *et al.*, 1996), the question of sustainable soil fertility management remains controversial, even though the assertion of runaway degradation seems to be at odds with the fact that rural populations have doubled in the last 30–40 years and still support themselves, to a very large extent, from agriculture.

In this chapter, we will first identify the physical and biogeographical context of soil fertility management in our four study areas, and describe the technologies and modes of management employed by farmers. We will then analyse intensification in empirical terms in four steps. First, we examine the relations between rural population densities and land use. Second, we analyse farm labour use per hectare under differing conditions of ecology (rainfall) from place to place, from year to year and from task to task. Third, we redefine the concept of farm labour use per hectare as *peak* labour use during key farming operations, when the labour constraints may be best observed at work. Fourth, we look at technological change, and in particular, the integration of crop with livestock production.

Working with the earth

The Hausa word *k'asa* has a range of meanings, from a handful of earth to a district or a country (a 'land'). Thus it can stand for a farmer's private right to the use of the soil, or for a sense of ethnic or even national rights to a territory. In a parallel sense, the actions of individuals in exploiting or conserving the soil on their farms, plots or parcels of land also range up to higher scales, and ultimately determine the sustainability of the Sahel as we know it.

Soils in this part of the Sahel are derived from wind-blown parent materials deposited during Quaternary arid phases, and are brown and sandy textured, with free drainage, poorly developed profiles, low clay fractions, deficiencies in phosphorus and (unless fixed by leguminous vegetation) nitrogen (Jones and Wild, 1975; Pieri, 1989). In river valleys and depressions, although usually less than 10 m below the upland in topographical terms and comprising less than 10 per cent of the area, darker soils, often having more clay and impeded drainage, are found; in places where drainage is restricted, as in Kaska, high saline or alkaline levels may occur. These areas are flooded or waterlogged relatively frequently (Turner, 1997).

However, within this general description there is much local diversity, and farmers in particular have learned to exploit it to their advantage. Soil conditions tend to vary not only between places but between neighbouring fields as small as one-tenth of a hectare, and even within fields, calling for a variety of soil management solutions. Farmers usually categorise soils by their colour, but recognise their moisture holding, crop suitability and other properties.

Soil management in Tumbau (Essiet, 1994; Harris, 1996; Yusuf, 1994, 1996)

In Tumbau, cultivated farmland includes both upland (*tudu*) and lowland (*fadama*). Our field investigations (using inventories, sampling and profile pits, a detailed study of fertility and its management) were confined to upland soils which comprise approximately 90 per cent of the 11 × 11 km block containing the rainfed Tumbau farmlands. The ecology and management of *fadama* is dominated by surface or subsurface water availability throughout the year. Stream courses grade downwards towards the Kano River, whose extensive flood plain is about 20 m below the upland, with a generally flat or very gently sloping surface. Under intensive rainfall, not only the lowland soils but also pockets of upland may be covered by standing water for short periods.

Sandy ferruginous soils (latosols) are lateritised on exposed slopes and more extensively below the surface. The weathered regolith of the Basement Complex is covered by aeolian drift and the soils are mature, well drained and acidic with high percentages of sand and low in organic matter, usually less than 0.5 per cent. According to Bennett *et al.* (1978), the soils of the Kano Close-Settled Zone are generally deeper than 100 cm, well drained, lacking a coarse material layer, and with iron segregation mottles occupying less than 20 per cent of any horizon. The topsoil and subsoil colour are variable (possibly owing to intensive cultivation), but the dominant B horizon colour is yellowish red.

Farmers in Tumbau recognise two upland soil types on the basis of colour. *Rairayi* (literally 'white one', actually light yellowish to yellowish brown) is lighter and retains less moisture. *Jangargari* (literally 'red one', reddish yellow or strong brown) is heavier, retains more moisture and gives better residual effects from fertiliser. There is no significant difference in either fertility or management.

Tumbau has supported a high population density for centuries. Smallholder farming is highly intensive and the farmlands are under continuous cultivation. In order to maintain fertility and yield, the system has developed diverse agricultural management practices in terms of cropping, fertilisation and productive technologies. Twelve distinct management practices based on fallowing, farming technologies, fertilisation and agronomic practices were identified in the area (Yusuf, 1996).

Soil fertility is chiefly maintained by applying different types of animal manure, notably small ruminant pen manure, cattle manure, domestic refuse, ash, and burnt gathered leaves and crop residues. Inorganic fertiliser is in high demand but only a few farmers can afford to purchase it in adequate quantities. It is largely applied to grain (sorghum and millet) when available and affordable. Generally, the application of some form of fertiliser is necessary for a reasonable harvest. Therefore, farmers do all they can to supply manure and fertilisers to their farms.

The most important source of manure is from penned domestic animals, especially small ruminants. There is no family in the area that does not keep some small ruminants, and their function in providing manure is second only to their role as a pool of savings. Those who have few livestock borrow small ruminants from relatives or friends so as to pen and feed them in order to obtain manure. Farmers feed the livestock with all sorts of leaves, grasses, crop residues and browse. Collection is time-consuming, especially in the last part of the rainy season (August). Domestic refuse and ash are thrown into the livestock pen. In the dry season when the livestock are released for free grazing, the farmers remove the manure to the farms on donkeys (about 75 per cent), ox-drawn cart (20 per cent) and by head, especially to home farms (5 per cent).

The farming system in the area ensures a cycle of nutrients. Manure is needed to produce crops, and crop residues are used in feeding the livestock which produce manure that is taken back to the farm to improve soil fertility. The cycle is fairly efficient, though some nutrients are lost by taking some harvested crops away from the area. Nevertheless, this is compensated by the browse and leaf litter from the trees that exploit nutrients from the bottom of the soil horizon, and also by the application of chemical fertilisers brought from outside the system. Nutrients are transferred to the three subsystems: livestock, crops and soils (Harris, 1996). Farmers know that the soil cannot produce a satisfactory yield without manure, and also that the survival of their livestock depends on the crop residues and farm weeds. Though there are a number of leakages in the cycle, and some inflow of nutrients from other sources, they are insignificant on a larger scale.

Hand ridging has been overtaken by ox-drawn ridging ploughs, which were introduced into the area at least 25 years ago but have become widely popular only since 1985. Because of the heavy capital investment involved, few families own ploughs (about one in 25). Yet these plough teams are sufficient to cultivate their own and other people's farms (on hiring), though there are queues and delays in the first three weeks of the rainy season for non-owners, and some may not have the money to hire in time. The prime ambition of most farmers in Tumbau is to possess an ox-plough as a means of saving labour, of maximising crop production and of earning income.

Soil management in Dagaceri (Mortimore, 1989; Essiet, 1994; Mohammed, 1996)

In Dagaceri, the parent material of the soils is deep, stabilised dune sand. Inter-dune depressions are sometimes aligned from ENE to WSW (the prevailing wind direction during the Quaternary desert transgression). The relative relief is of the order of 10 m. The dune soils weather to a reddish colour on some crests and are white elsewhere. There are no wetland areas, though flooding may persist for a day or two after heavy rain in the interdunes (which have no integrated drainage). A few seasonal ponds are associated with areas of black or grey soils.

Farmers recognise three main soil types, whose names are in the Manga language. *Kati-kime* (literally 'red', actually yellowish red), has incomplete mineralisation of potassium feldspar, weak angular blocky structures, sandy texture, micropores and is generally porous. It occurs near the tops of dunes. *Keza-keza*, (literally 'white', but a strong brown below the surface) has a weak blocky structure, surface capping, micropores and sandy texture. This is the most common farmland soil. *Tulo-tulo* (literally 'black', but dark brown below the surface) is hard and compact with massive and crumb structures, a sandy clay texture, mottles, and both macro- and micro-pores. This is found in depressions. Soil pits revealed only a limited differentiation of horizons in all three types. Farmers plant all crops in all soils, though with closer spacing in *tulo-tulo*. Many of them farm only on *keza-keza*, which accounts for perhaps 70 per cent of the surface.

Farmland in Dagaceri consists of manured infields (fertilised with livestock dung, ash heaps and compound sweepings), and unmanured outfields. Manured fields are seldom fallowed. Cultivation for 25, 30 and even 50 years is reported on manured farmlands close to the village. Unmanured outfields are fallowed, but unsystematically, as farmers lack the input and labour to cultivate them. Farms in interdunal depressions where water collects, grow more weeds and need more labour than those on dunes. Manure or fertiliser also encourages the growth of weeds, so these fields too need more labour.

Ploughs were first introduced to the area when groundnuts were grown for export in the 1960s, but all the bulls were lost during the Sahel drought of 1972–1974. During the 1980s the plough was reintroduced, and now half or more of the fields are thus cultivated. It is used mainly for ridging before sowing, and confers quicker planting and more effective germination than planting in the flat, which, however, can be done before the rains begin (*binne*) which gives a speedier response to early rain. The poor man's alternative for saving labour is the *ashasha*, the long-handled weeding tool with a crescentic blade, which is pushed forward just under the soil surface rather than being pulled backwards and deeper, as with the hand hoe (*fartanya*). The *ashasha* is twice as fast as the hand hoe.

Soil management in Futchimiram (Chiroma, 1996)

Only one soil type is recognised in Futchimiram, which has a gently undulating surface (a relative relief of less than 6 m) covered by open woodland and annual grasses and herbs, with a few very episodic shallow pools. The parent material is deep Chad Formation deposits of alternating coarse and fine sands. The soils are sandy and porous throughout the profile, low in organic matter, total nitrogen, available phosphorus and exchangeable bases.

When the woodland was cleared around a settlement, strip fields were established from each household, mainly westwards. These fields, known as *kulo fatoye* (Kanuri: home farms), being those of the first settlers, have been under continuous cultivation ever since, without fallow. In the hamlet studied, this means at least 40 years. The fields of a household normally include one large one devoted to millet, and a few smaller ones devoted to groundnut, cowpea and sorghum. Repeated monocropping of rainfed crops is common with virtually no intercropping. Rotation and fallow, however, are introduced whenever soil fertility drops too low. Annual cultivation of the same crop for ten to fifteen years is common on newly opened fertile fields, or on home farms enriched with organic manure. In the absence of sufficient manure for all their fields, farmers prefer to fertilise the home fields and to rely on short fallowing to maintain the fertility of more distant outfields. Each household has its cattle alongside the house, and often stations them on the nearby home farms for manuring in the dry season. Rotation is usually practised between cereal crops (millet and sorghum) and legumes (groundnut and cowpea), although the frequency of rotation varies.

These activities occur within the life cycle of a settlement which, in the past, would move to a new site and transfer its cultivation to newly opened woodlands after a few decades. Although such movements are still possible, they are increasingly constrained by the numbers of other settlements established in the area, and shifting cultivation is usually confined now to the collective periodic opening (and abandonment) of blocks of individually farmed fields within range of the existing hamlet.

Few ploughs are yet in use in Futchimiram (they were first introduced under a credit scheme operated by the North East Arid Zone Development Programme in the 1990s), and the *ashasha* is not used. Cultivation is therefore dependent on the use of the hand hoe. Ridges, however, are rarely made, as the soil drains freely.

Soil management in Kaska (Mortimore, 1989; Essiet, 1994; Ibrahim, 1996)

In Kaska, the relative relief is greater than in the other areas, with about 20 m separating the upland (stabilised or moving dunes) from the bottomlands of the steep-sided, closed circular or elliptical depressions which often contain saline ponds. Intermediate in terms of relief are sandy-floored lowlands, thought to be infilled depressions. This area had an active dunefield until relatively recently and has still not developed a woodland vegetation, except in the depressions where dense woodland takes advantage of shallow groundwater. Flooding occurs in the deep depressions after heavy rain.

Farmers follow these topographical elements in their classification of the soils (for which the Hausa terms are used here):

* *tudu*, upland, yellow or white dune sands lacking any profile, very freely draining, but sometimes differentiated by colour within a short vertical or horizontal distance, disturbed by frequent remobilisation in the recent past, and supporting only annual grasses;

- *faya*, intermediate lowland, white sands, freely draining, but not subject to remobilisation, supporting grasses, shrubs and scattered trees;
- *kwari*, deep depression alkaline soils with silt and clay present in significant quantities, hard surface capping, rainwater running off into the ponds, supporting (except where capped) dense woodland, with groundwater from one to three metres below the surface.

Sharp differences in management reflect the same categories. Until the 1980s, almost all farming was confined to the *faya* and some *kwari* soils, the latter being used for irrigated farming during the dry season where groundwater is less than 2 m deep. The *kwari* soils are recognised as the most fertile, unless impeded drainage, salinity, alkalinity or surface capping are problematic. The *tudu* soils were left for grazing. With reduced rainfall, the dried out *kwari* soils are hard to cultivate for rainfed farming. This activity has since been extended from the *faya* to an increasing area of *tudu* soils. Most farmers have access to all three soil types, the *faya* still providing the basis for rainfed farming, and the *kwari* being restricted to dry season cultivation.

The interdune (*faya*) depressions are now the most intensively cultivated, apart from the irrigated (Kanuri: *garka*) plots in the *kwari*. Their soil is sandy, mixed with a little clay. It has a high moisture retention capacity but is easily managed compared with *kwari* clay soils. Fertility is generally low, and soils cannot withstand more than five years of cultivation. The Manga farmers maintain soil fertility by short fallows (two to five years), as they have insufficient livestock to manure them adequately, though they invite the pastoralists in their neighbouring settlements to fertilise the soil through livestock droppings in exchange for crop residues or friendship. Under this system, the *faya* fallow duration is usually between two and fifteen years, depending upon the labour available, the needs of the household, the capacity of the soil and drought. Under the fallow system, the Manga grow millet and sorghum (usually for subsistence, although when there is a surplus or a social or health problem they are sold), *guna* melon and a little benniseed. Currently, there is no *faya* land that is not under cultivation or has no title attached to it.

The FulBe agropastoralists also farm *faya* soils. FulBe farmers practise a semi-permanent system of cultivation with no fallow at all, or with fallow at long, irregular intervals (e.g. during the droughts of 1983 and 1985 when some FulBe migrated temporarily to wetland areas). Organic manure obtained from their livestock sustains long periods of cultivation. One farmer's land has been under cultivation for over 50 years without fallow. The FulBe cultivate mainly the subsistence crops, millet and sorghum. They do not participate in commercial crop production, cowpea or *guna* cultivation, and they have no *garka* plots in the *kwari*.

Owing to the scarcity of *faya* land, and the failure of the *kwari* farmlands, Manga farmers are now extending their cultivation to the slopes and tops of sand-dunes. This practice started during the *Banga-banga* drought (1983–1985).

Millet, sorghum and *guna* are grown. Farmers cultivating *tudu* areas are facing the problem of crop damage by the livestock, for as most of the *tudu* areas are unappropriated, animals graze there. Animals wander into the tudu farms and destroy the growing crops, thus resulting in conflicts between the Manga and the FulBe.

Neither the Manga nor the FulBe farmers use ploughs, even though the latter have access to their own bulls. The *ashasha* is used for weeding work. Fields are not systematically ridged. However, in the irrigated *garka* plots, the Manga farmers pay close attention to micro-catchments, bunds, channels and other agronomic responses to the soil and water conditions.

Fertility management

Each of the four farming systems has a regime of fertility management which reflects the special conditions of soil, water, labour and need that characterise its neighbourhood. No farmers use inorganic fertilisers on a regular and sustained basis. This is, by universal lament, on account of their cost and supply limitations. Thus the better off are the most frequent users. Market crops tend to receive applications in preference to the subsistence grains. Three farmers in Tumbau, in the years 1993 and 1994, used an average of 33 and 37 kg per hectare of NPK fertiliser, compared with 3.3 and 4.7 tons of organic manure (Harris and Bache, 1995). In Dagaceri, fewer farmers use less inorganic fertiliser on fewer of their fields. In Futchimiram and Kaska, fertiliser supply is completely dependent on the development programme, and in the longer term it has been negligible.

We shall examine intensification, therefore, not as a technical option for the future, an outcome of developmental interventions, but as an indigenous process whose historical path can be reconstructed from longitudinal indicators, and whose contemporary status can be approached in empirical terms. This provides a more realistic basis for evaluating its potential in future.

Population density and land use

Boserup's (1965) now classic statement of intensification identifies a link between an increase in rural population density and land use, or specifically, the frequency of cultivation. As land becomes scarce, cultivation cycles become longer relative to fallows, which means that each hectare of cultivable land is cultivated in more years out of every ten. According to Young *et al.* (1981), shifting cultivation systems in the tropics achieve a maximum of about two years in ten, when approximately 20 per cent of cultivable land is in use at any one time. With higher population densities, this fraction increases and as fallows become shorter, more land is put into annual or permanent cultivation on fixed fields. When it reaches the limit of the cultivable land available, shifting cultivation is abandoned.[1]

Table 6.1 Rural population densities and land use in four study areas

	Tumbau	Dagaceri	Futchimiram	Kaska
District population density, persons/ km² (1)	414	66	13	18
Local population density, persons/ km²(2)	223	43	31	11
Productive land, per cent (3)	89.5	88.8	91.7	78.0
Arable land, per cent of productive land	98.8	61.5	23.7	13.7
Arable land, ha/person (4)	0.46	1.9	1.29	1.52

Sources: Federal Republic of Nigeria (1992); Turner (1997); cadastral surveys.

Notes
(1) Census of 1991 (figures for Gezawa, Birniwa, Geidam and Yusufari local government areas). These include urban population, but the figure for Geidam LGA (Futchimiram) is claimed to be a major underestimate.
(2) Estimated densities for blocks of approximately 120 km² derived from air photo interpretation, dated 1981 (Tumbau and Dagaceri) and 1990 (Futchimiram and Kaska), in four blocks of approximately 120 km². Controlled by an enumeration of 50 per cent of the population in Dagaceri, and of collaborating households in the other areas.
(3) Interpreted from air photographs. Productive land is that which is available after subtracting degraded land, active dunes, water or lake beds and settlements.
(4) Field surveys of collaborating households numbering twelve in Tumbau, thirteen in Dagaceri, six in Futchimiram and fourteen in Kaska.

In general terms, the hypothesis receives support from numerous studies at village or district level in the Sahel. The crucial issue is whether, having abandoned shifting cultivation on some or all of its land, a farming system can find alternative ways of maintaining bioproductivity. Fallowing is reduced to one or two years which is known from soil analyses to be inadequate. Considerable importance, therefore, is attached to systems which do maintain themselves at relatively high densities (Turner *et al.*, 1993). Are these sustainable, and if so, do they offer a path for other, transitional systems?

In Table 6.1, an association is apparent between local population densities and the percentage of land that is arable along our transect of four sites. The term 'arable' includes both land under cultivation and short fallows (usually of one or two years) because, in air photographs, it is often impossible to distinguish between grass fallows and poorly weeded cropland. However, while the trends are parallel, the values do not match well in ratio or proportional terms. The relationship between population density and arable land per person is far from perfect. In particular, Dagaceri (at 1.9 ha) is out of line with the others.

Does this broad association, which results from a comparison between places at similar points in time, mean that as population density increases over time in the same place, the arable fraction must also increase (Table 6.2)?

Historical rates of population increase in the Sahel have been put at 2.2 per cent per year (IUCN, 1989). Census data are unreliable in Nigeria. However, according to official sources, the district population density (including both

Table 6.2 Changes in the proportion of arable land (as a percentage of all land): four study areas

		1950[1]	1969[2]	1981/90[3]
Tumbau	upland	75.8	87.2	87.7
	lowland	1.8	2.1	0.7
	total	77.6	89.3	88.4
Dagaceri	upland only	35.6	56.1	54.6
Futchimiram	upland only	22.1	22.5	21.7
Kaska	upland	0.9	0.4	10.0
	lowland	14.9	3.8	0.7
	total	15.8	4.2	10.7

Source: Turner (1997).

Notes
1 Futchimiram: 1957.
2 Tumbau: 1971.
3 Tumbau and Dagaceri: 1981; Futchimiram and Kaska: 1990.
  Areas mapped: 102–115 km^2 (Tumbau); 132–146 km^2 (Dagaceri); 135–170 km^2 (Futchimiram); 129–146 km^2 (Kaska). The differences are due to air photo availability.

urban and rural populations) of Tumbau increased from 149/km^2 in 1931 to 192/km^2 in 1952, 226/km^2 in 1962 and 414/km^2 in 1991. That of Dagaceri increased from a mere 17/km^2 in 1931 to 29/km^2 in 1952, 47/km^2 in 1962, and 66/km^2 in 1991. Of a long-term upward trend in densities there is no doubt. However, the profiles of land use change after 1950 in the four villages lend little support to a model driven *in detail* by demographic change.[2] In the event, each study area has pursued a unique pathway, which has been determined by a combination of general and specific factors:

- In *Tumbau*, our period began (in 1950) at the end of an historical process of landscape transformation in which natural vegetation, including even short fallows, was almost completely eliminated in favour of annual cultivation in permanent fields. The arable fraction could not increase significantly, no matter how large the population; it is no longer a significant indicator of continuing intensification in the farming system.
- In *Dagaceri*, an annual increase of 2.7 per cent in the population density (1931–1952) was accompanied by an annual increase of 2.4 per cent in the arable fraction (1950–1969): a predictable outcome of the general model. However, during that period, the groundnut boom in northern Nigeria induced almost all farmers to expand their cultivation, but after 1969 the expansion of the arable fraction stopped – while the population density went on increasing. Even today, farm holdings in Dagaceri (at 1.9 ha/ person) are the largest among our four villages (Table 6.1). These two anomalies can be explained in the following terms. In 1972, most of the remaining unclaimed land was reserved for grazing. In anticipation of this

event, which resolved long-standing competition between Manga and FulBe ethnic communities in the district, the Manga farmers had an interest in maximising their claims to available arable land, and securing those claims by keeping it under cultivation. After that the arable fraction could increase no more.

- In *Futchimiram*, a remarkably stable system of shifting cultivation has been maintained over four decades or longer. This system uses a relatively high input of labour per hectare, partly to cultivate commercial groundnuts at high stand densities. Such stability suggests a priori a slow increase in the population density – or none.
- In *Kaska*, a complex (Manga) farming system combines rainfed farming with perennial cultivation (sometimes irrigated) in lowland sites. There were wide fluctuations in the arable fraction, which were the result of shifting *faya* cultivation moving in and out of the block of land that was mapped. After 1969, there was a shift away from deep depression (*kwari*) farming in favour of upland (*tudu*): a change having little to do with the demographic trend.

Brief though this exposition of land use changes is, many contributory factors are already identified. These include population growth, land availability, markets, changes in the rules of access to land, shifting cultivation, crop choice and rainfall decline. It cannot even be said that, among such a diversity of factors, population growth always plays a dominant role. But in Tumbau and Dagaceri there is no more unclaimed land available for cultivation. This situation has been termed 'saturation' in the francophone countries of West Africa.

We therefore end this discussion of the relations between population density and land use with a qualification and a warning. The qualification is that the arable fraction may be a poor indicator of intensification. The warning is that a general model – even when superficial analysis appears to verify it (as in Dagaceri before 1969) – requires confirmation on the basis of local, long-term investigations.

Total labour use

Land use transformation is a visible end-product of labour (and technology) applied to the land. The pathway of intensification, as identified in 'Boserupian' interpretations, involves increasing inputs of labour per hectare. The imperfections of using land use as a proxy indicator of intensification are apparent in the preceding section, so we proceed now to grapple directly with labour use in our four villages.

Labour use per hectare

The methods used in our study of labour (and the methodological pitfalls) have been described in Chapter 2. Table 6.3 sets out the amounts of agricultural labour

Table 6.3 Labour use per hectare in one farming year (1996)[1]

	Tumbau	*Dagaceri*	*Futchimiram*	*Kaska*
Agricultural labour used (units/ productive ha)[2]	67.7	11.3	14.6	3.7
Farm labour used (units/arable ha)[3]	50.4	13.3	36.4	26.4

Notes

1 The data are seasonally adjusted; as the farming season is longest in Tumbau and shortest in Kaska, a standard period is used (from the second week of June to the last week of November).

2 Agricultural labour is farm labour plus livestock labour. Productive land is that which is available after subtracting degraded land, active dunes, water or lake beds and settlements.

3 Farm labour was all spent on the arable fields used by the collaborating households in 1996.

used per productive hectare, and the amounts of farm labour used per arable hectare, in the year 1996.

The use of agricultural labour (that is, farming and livestock labour) per productive hectare conforms closely with an expectation that a regular function should link the demographic resource with the amount of labour expended (Table 6.1).[3] However, such a relationship is not sustained when we measure farm labour use (deducting livestock work) per arable hectare. The reason why labour inputs per hectare fall sharply in Tumbau, while rising in Futchimiram and Kaska, when livestock work is deducted, is that in the Kano Close-Settled Zone, nearly all land is arable, livestock and crop husbandry are integrated (Hendy, 1977; Harris, 1996), and animals consume a lot of time. In Futchimiram and Kaska, farm labour is concentrated on a small arable fraction. It is clear from the table that it would be misleading to exclude livestock work from an assessment of intensification.

Livestock play an important role in all four systems, but their husbandry is more integrated with crop production in Tumbau than in any other. The criteria we use for measuring integration are: (1) the use of crop residues as fodder (relative to natural rangeland); (2) the making and distribution of animal manure on farms; and (3) the use of animal energy on farms (ploughs, weeders or carts). On all three, Tumbau ranks first. However, what is particularly relevant to the figures shown in Table 6.3 is the extent of territorial integration; that is, the convergence of crop and animal husbandry within the boundary of the private farm. Whereas in the lower intensity systems livestock depend to a significant extent on forage derived from rangeland or fallow, in Tumbau, where such land is extremely scarce, they are supported almost entirely from residues and weeds grown on cultivated fields.

Livestock work follows a low-fluctuating curve through the farming season (as animals' requirements do not vary much from day to day), compared with the high fluctuations in farm work which are associated with peaks in planting, weeding and harvesting (Figure 6.1). But whereas in Tumbau the livestock commitments of twelve households are high, in the other villages those of

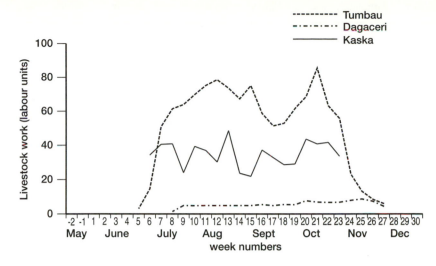

Figure 6.1 Livestock work in three villages, 1995. (Work is shown in weighted units (equivalent to a man-day) per seven-day period.)

thirteen households are low by comparison. In Tumbau much labour is used in collecting fodder and carrying it to the animals, as they are penned or tethered day and night, household by household. Tending animals in the other villages means supervising grazing on rangeland and fallows, normally done in Kaska by a small child, household by household, or in Dagaceri, put out by households to a village shepherd. When the harvest is over, labour spent on tending animals at Tumbau falls rapidly, as they are allowed into farmland to graze unsupervised; but shepherding is still required in Dagaceri and Kaska, to prevent the loss of animals in the open rangeland.

Labour use by task

However, leaving livestock work aside, the core of the intensification process is found in the use of farm labour on arable land. In Table 6.4, this is broken down according to the major farm tasks.

The table shows once again that the Tumbau system is in a class of its own when compared with the other three lower intensity systems. This is expressed in the following ways.

Land clearing and fertilising

Much higher labour inputs are required to sustain the practice of manuring, which is carried out during the growth cycle of the crops on a plant-by-plant basis. In Tumbau there is usually little clearing work to be done, as all fields are

109

Table 6.4 Farm labour use (in weighted units) per cultivated hectare in one year (1996)[1]

	Tumbau	*Dagaceri*	*Futchimiram*	*Kaska*
Hectares[2]	45.69	116.4	25.7	132.2
Average ha/capita	0.35	1.6	1.4	1.3
Average ha/labour unit available	0.7	4.0	2.5	2.6
Land clearing and fertilising	4.76	0.98	na	0.11
Planting[3]	5.51	3.14	1.41	2.24
Weeding and thinning	31.55	8.38	21.35	16.20
Harvesting	35.11	3.99	11.19	3.51
All farming	76.89	16.48	33.94	22.07

Notes

1   The data are not seasonally adjusted (cf. Table 6.2), because by doing so, some major tasks would be partly removed from the record in Tumbau, where the farming season is longer. Periods of monitoring are: Tumbau, week numbers 3 May to 4 December; Dagaceri, 1 June to 3 December; Kaska, 2 June to 4 November; Futchimiram, 4 May to 3 November.

2   The numbers of hectares given in this table are the cultivated and short-fallowed fields used by the collaborating households, omitting those for which either area or labour use values are unreliable (and so may differ from those given in Table 6.2).

3   Planting labour was incompletely recorded if (a) dry planting occurred (before the rains began), which was common in Dagaceri, Kaska and Futchimiram, or (b) the 'planting rains' arrived before the commencement of monitoring (up to one week was lost in Tumbau and Kaska in 1996).

na Not available.

cultivated every year, and weeding and the removal of residues in the previous year were efficiently done. Additional labour was used (before the start of monitoring) to carry manure from the animal pens and compounds to the fields, where it was stored in scattered heaps until the rains began. There is some manuring in Dagaceri; but there, as in the other three villages, the category 'clearing and fertilising' mainly reflected uprooting or cutting down vegetation which was regenerating naturally in the fields. It might be done quite late.

Planting

Higher inputs in planting resulted from the higher planting densities used in Tumbau, and the practice of multiple cropping, which sustains planting activity for several months. Late planting of sorghum also occurs in Dagaceri, and *guna* (which is planted after early millet has been harvested) also raises planting labour there.

Weeding and thinning

Higher inputs in weeding and thinning reflect a larger number of weeding cycles in Tumbau – three or even four against two or even one and occasionally none (owing to labour scarcities) in Dagaceri and Kaska. The heavy inputs of labour in weeding (which until recently was done entirely by hand) are a necessary condition of relatively high yields – a classic example of labour intensification.

Harvesting

Much higher labour inputs in harvesting result from the higher output of biomass in Tumbau – every bit of it having economic value – and the extended harvesting operations entailed by multiple cropping. Harvesting actually consumed more labour – in total – than weeding and thinning, which are usually considered to be the most constraining demands on the labour force.

However, if the most densely populated area is found, as expected, to support the highest labour inputs per hectare, the data for the three lower intensity systems fail to show a regular decline with diminishing population density. Their eccentric patterns are explained, again, in terms of factors which are specific to each place.

Dagaceri

The local administration's reservation of grazing areas in Dagaceri for the FulBe (to whom cattle herding, though not always ownership, is restricted), referred to above, put a stop to the expansion of arable land held privately by Manga farmers. Prior to this, and possibly anticipating such a restraint, farmers had been appropriating large areas, which were subsequently subdivided among their heirs. As free land was no longer available for shifting cultivation, fallowing (and thus the frontier of cultivation) became internalised within the boundaries of private farms. Fallowing is not used necessarily as a strategy to enhance soil fertility, but is as often justified as an inability to mobilise enough labour for its cultivation (and, if necessary, fertilisation). This operational concept of fallow may explain why Manga farmers frequently plant larger areas than they can later weed, having to abandon millet after one weeding, or even none at all. Such abandoned fields (or parts of fields), heavily overgrown with grass, cannot easily be distinguished from fallows. This accounts for the facts that planting labour per hectare was quite high in Dagaceri, but weeding labour was relatively low.[4]

Futchimiram

Shifting cultivation is still practised in Futchimiram, not, by and large, by individuals, but by groups of farmers. This is because the Badowoi keep large numbers of cattle and small ruminants and they continue to graze the rangelands (which are, in reality, long fallows) during the cropping season. Livestock are kept off the crops by means of thorn fences constructed from branchwood, which is obtained by pollarding mature trees found standing on the farmland when it is brought into cultivation. Such fencing is more economical if fields are grouped. Furthermore, the permission of the village head (*Zanna*) of Futchimiram is necessary before a new clearing is made. This permission, which is given to the farmers' groups, is dependent on vegetational indicators of the recuperation of soil fertility since the last cycle of cultivation. Thus, in this system, fallowing is largely externalised (although it sometimes occurs on large, private fields, and is usually due to labour shortages). This is one reason why labour use per enclosed hectare on weeding/thinning and harvesting is relatively high in Futchimiram.

Another difference between the Dagaceri and Futchimiram systems is that plant spacing is far wider in Dagaceri fields – a metre or more separating stands of millet – than in Futchimiram fields, where stands are commonly about 0.5 m apart. Furthermore, groundnuts are grown in Futchimiram in small patches at very high densities in order to minimise weed growth. These patches can be moved around the farm from year to year. In Dagaceri, groundnuts have failed since 1975. Thus the system of Futchimiram can be characterised as a 'mobile intensive system' whereas that of Dagaceri is a 'stationary extensive system'.

Kaska

Kaska appears to be closer to the Dagaceri than to the Futchimiram model (Table 6.4). This, however, hides divergences in practice which reflect controlling topographical and soil factors. Shifting cultivation (associated with periodically shifting settlements) continued until recently in the preferred *faya* depressions, according to air photographic evidence, on land which it is now unwise to vacate for fear of losing claims. Valley bottom (*kwari*) land, on the other hand, is sometimes abandoned for rainfed farming, though if within reach of irrigation water, it supports intensive labour use outside the rainfed agricultural year. Upland (*tudu*) land, little used for farming before the 1980s, is now subjected to a rapid increase in private appropriation.

These findings indicate that the use of total labour inputs per hectare of arable land can be a misleading guide to intensity unless the data are placed in their systemic context. Our transect has exposed some of the diversity which exists and some of the dangers in attempting comparisons. Is there a more satisfactory indicator of labour intensification?

Peak labour use

Efficient management of labour is most important when it is most scarce. In order to eliminate the distorting effects of low-intensity activities before and after the growing period, and improve compatibility between growing seasons of differing length, in this alternative measure of labour intensity we restrict ourselves to peak labour inputs during major operations. These are (1) weeding and thinning, and (2) harvesting. Of these, the first is the more critical, as everyone agrees that crop output is determined first by rainfall, and second by weeding efficiency.[5] The peak labour input is defined as the average of the five highest weeks in each task.

Labour use in major farming operations is influenced by conditions of rainfall and growth in a given year. Thus, after a hesitant start to the rains, planting may have to be repeated several times, increasing the amount of labour used (though not yet at the expense of other tasks). A short drought during the growing season reduces the requirement for weeding, as weeds fare worse than well-rooted crops. A longer drought reduces the harvest of both crops and residues. In this way, ecology sets ceilings to labour intensification, not in this case through average conditions (as is generally understood), but by its variability. This mechanism generally works through rainfall, but other ecological agents can easily be recognised (for example, insect pests, crop diseases). Ecological capping of intensification through sub-optimal conditions can thus occur during the short window of opportunity which opens in the growing period. In this way, average labour use per hectare is kept down, though in a good year it may rise relatively rapidly. It is its ability to respond to such an opportunity that really defines how far a system can intensify under the variable conditions of the Sahel.

The extent of such variability is shown in Figure 6.2, which charts peak labour inputs in the two major tasks of weeding/thinning and harvesting over the four years 1993–1996, and for each village (though data for Futchimiram are only available for 1996).

As each value represents the average total labour input by all the collaborating households, on all of their arable land, during one seven-day period (the average of five peak periods), quite small differences are operationally significant. The differences between average labour use per hectare in Tumbau, Dagaceri and Kaska are broadly consistent with the hectares of arable land available per labour unit; the more land there is, the more thinly labour effort must be spread.

The values are of some interest in absolute terms. Even on the small holdings of Tumbau (with 0.7 ha available per labour unit), the equivalent of three man-days per hectare appears to be a maximum attained on either weeding/thinning or harvesting activities. Either greater inputs encounter negative returns to effort, or constraints such as energy, fitness, working time consumed by farm fragmentation or the call of other commitments (among which we know that livestock are pre-eminent), must restrict higher inputs. On the much larger holdings of Dagaceri (with 4 ha available per labour unit), the highest average inputs rise to just over one man-day and the lowest fall to just over half (in weeding) and one-third (in harvesting).

Inter-year variability in the peak intensity of labour use is relatively less in Tumbau and greater in Dagaceri, though greatest of all in harvesting in Kaska. Intensity, measured by this indicator, suggests that Tumbau benefits from a superior agroecological potential, as shown by its average rainfall (Figure 6.2),

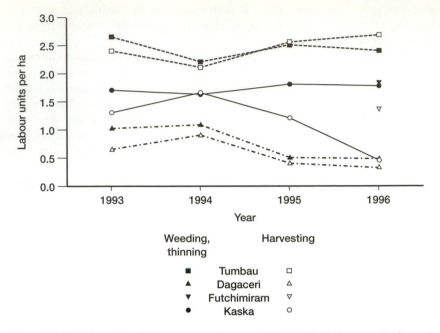

Figure 6.2 Peak intensity of labour use per hectare, 1993–1996. (Labour use is shown in weighted units (equivalent to a man-day) per seven-day period.)

but that among the three remaining systems, intensity is unrelated to rainfall. This was to be expected, in view of our emerging characterisation of the Tumbau system as relatively invariable, inflexible, complex and intensive, and of the Dagaceri system as variable, highly flexible, less complex and extensive.

Inter-year variability in the use of labour tends to be greater in harvesting work than in weeding/thinning. This was to be expected, as the economic value of the harvest (crops and residues) determines the amount of labour time spent in gathering it in, and the rainfall (which is *not* under the farmer's control) is the primary determinant of biomass output. Earlier in the season during weeding/thinning, outcomes are not known, and a primary determinant of production (which *is* under the farmer's control) is weeding efficiency.

Interpreting intensity

In this chapter we have attempted to insert the technologies of soil fertility management, and the management of agricultural labour by farming households, into the matrix of intensification as an indigenous and adaptive process. We began by describing the soils and their management in each of the four study areas. Then, in approaching intensification from the conventional angle of rural population densities, we asked whether, in these areas, a land use indicator (the cultivated fraction) is positively related to population density, either between places or over time. While confirming the appearance of a spatial relationship we found no clear trend through time, and that a simple model is severely restricted both by the limitations of the indicator itself and by the diversity of particular situations. These findings expose some of the complexity of evaluating intensification and thereby relating human labour with soil productivity.

As the limitations of discussing intensification in terms of population density and land use are obvious, we approached the measurement of labour use per hectare directly. We find the predicted relationship with population density when agricultural labour inputs are averaged over all productive land, but not when farm labour inputs on arable land are considered. While the high intensity of the Tumbau system is clearly beyond doubt, in the lower intensity systems the use of total labour inputs per hectare of arable land can be a misleading guide to intensity unless the peculiarities of each system are taken into account. At this stage, we abandoned the integrated measurement of livestock with farm work (agricultural labour), concentrating only on farm operations (farm labour).

The poor compatibility of total labour per hectare, measured in various ways, instead favours peak labour per hectare used in major farm tasks – weeding/ thinning and harvesting. The analysis shows how the use of this indicator can facilitate comparisons both between different systems and between years. It also clarifies the interactions of labour intensification with rainfall variability and other ecological controls. The price, however, of improved compatibility is jettisoning not only livestock work but also other farm tasks.

A simple indicator continues to elude us. However, the simplification of debates about labour use has itself been unhelpful in the past. Dismantling the structure of labour use under differing conditions of agricultural intensity reaffirms the diversity and complexity of a process that has far too often been the subject of abstract generalisation. The sources of this diversity and complexity are several. They include the differences between places in their resource endowments, and the variability of rainfall between years. They also include fundamental difficulties with defining and operating suitable measures of the parameters of intensity. Given these complexities, it is clear that the term *intensification*, when reduced to the farm level, covers more than one meaning. All, however, are encompassed by the systemic drive to sustain productivity from finite land resources.

It is not possible to decouple intensification entirely from rainfall. A comparison between Tumbau and the other three systems (Figure 6.2) suggests a link which accords well with the known facts: a longer season allows more crops to be grown in a more intricate pattern, and the production of more plant biomass per hectare.

Given the variability between places and between years, and the variability between households (which has not been covered in this chapter), it is not

115

surprising that the decisions made by smallholders do not always fit into a mould defined for them by overarching theory. These decisions are also influenced by constraints and opportunities outside agriculture. We turn to these in the next chapter.

7

WHEN FARMERS ARE
NOT FARMING

Farming alone is not a way of life

Outsiders who have never attempted a subsistence livelihood can too easily assume that producing crops is the sole preoccupation of African farming households. Were this so, rural populations would have long since lost their foothold in the Sahel. The archaeological and historical record provides evidence enough of specialist potters, ironworkers, fishers and weavers from the earliest days of farming, as well as of early livestock keepers. Furthermore, even cursory attention to life in an African village can demonstrate a broad range of economic activities which are engaged in, and which are necessary to livelihood security. Yet too often this dimension has been ignored. A consequence of this neglect is that the long dry season seems to be evidence of idle hands and wasted labour (just as the diverse crop inventory and complex cropping patterns of wet season fields can suggest a dilettante disregard for the serious business of crop production).

Planners proposing interventions, such as new technological packages or irrigation schemes, have tended to assume that increasing the output of crops is the only available path to economic growth. Given such a premise, they may go on to assume that farming can command up to 100 per cent of available labour and will always be given priority over non-farming tasks, even where the intervention substantially raises the total commitment expected. The hostility of irrigation authorities to livestock is also well known. A failure of households to commit labour on the scale expected must then be put down to 'social factors' – which are, of course, beyond the remit of the technical planners. Such a standpoint reflects a single-sector approach to rural development, which was traditionally dominated by agriculturalists. An example was the Kano River Irrigation Project in northern Nigeria (Baba, 1974; Wallace, 1980, 1981; Palmer-Jones, 1984), and its successors at Bakolori and South Chad (Adams, 1991, 1992).

Labour supply is not constrained only by its demographic availability, and just as bottlenecks in the farming season are recognised to be a significant constraint on farm productivity, so also periods of the year when farm labour is not fully stretched should not be assumed to be times when it is 'idle'. It will not have escaped notice (Chapter 4) that the percentage of available labour used in farming, even in the peak periods of planting, weeding and harvesting, rarely exceeds 70 per cent and is usually much lower (Figure 4.5). The average in Tumbau is in the order of 30 per cent, and those of the other villages are only about 10 per cent higher.

Such assumptions can arise from three forms of neglect: (1) of the total context of household labour use, including a range of activities that have use value, in addition to the production of crops (or, for that matter, of livestock products); (2) of cultural factors that constrain the supply of labour, including gender roles; and (3) of the biophysical constraints on labour supply, such as sickness or other incapacity, age and energy. In this chapter we shall deal with the first of these; the second and third will be discussed in the following chapters.

Farming as livelihood

Every farming household must resolve the demands for non-farm labour, whose opportunity costs and social benefits must be weighed against those of farm work. The analysis of labour allocation which was carried out in our four villages uses six major task categories:

1 farm (crop production) work;
2 livestock work;
3 business activity (manufacturing, or selling good and services);
4 other off-farm work, including
   • work on other farms
   • time spent outside the village, e.g. at market
   • circulation to places more than a day's journey away;
5 domestic work (which includes fuel collection and some food gathering activities);
6 resting and sickness.

This categorisation represents a sequence in which the economic distance (and sometimes the spatial distance too) increases stepwise from the core activity of crop production. Categories 1 to 4 also represent a sequence of intensifying engagement with markets, in the sense that while purely subsistence crop production is conceivable (it is rarely practised by households, but not unusual for individuals, especially women), business and off-farm activity are inconceivable except in a context of exchange. For reasons which reflect the priority that Sahelian smallholders give to food production, it is rare for farming households to leave subsistence behind them and commit themselves fully to market production as do their large-scale counterparts elsewhere. So a household's strategic pathway from categories 1 to 4 is a journey of economic diversification, not of specialisation. The point of departure for this journey is never abandoned as staged evolutionary models of economic development imply. A better analogue of the process is *multiple stranding*, such as adding additional strands of *kaba* (palm) fibre to a string in order to make a thicker and more enduring rope.

The colonial powers opened up many land-locked African hinterlands, including those of the Sahel, to the power of world markets for primary products, and in many countries they deliberately assisted the penetration of these markets by imposing taxes, payable in cash (earlier taxes had been paid in kind), and by developing transport and market infrastructures and institutions, which distributed imported goods and fired rising expectations and patterns of monetisation in communities previously more oriented to subsistence. According to some interpretations, through competition among many small, unorganised producers and progressive lowering of the real prices for export crops, rural labour was devalued and an increasing dependency on markets for the satisfaction of basic needs was created. Cultivated areas were extended far beyond the needs of household food production. Yields declined, relative both to the areas planted and to the labour used. Regional migration to temporary or permanent employment in coastal towns, plantations or mines supplemented household incomes, but broke down social structures as individuals sought to privatise wealth. The loosening of ties to the land and to the community, and the concentration of wealth (including land and livestock) in the hands of trading elites, would, it was predicted, threaten to replace the peasantry with a proletariat. Famine crises from time to time served to deprive poor households of their remaining assets – even their rights to land where it was scarce. There is a large literature on the various nuances of this theme (for example, Rodney, 1972; Amin, 1974; Meillasoux, 1974; Raynaut, 1977b; Apeldoorn, 1981; Copans, 1983; Watts, 1983).

In the light of such interpretations, the sequence given above is regressive in its effects on household welfare in the longer term. As such, it must only weaken households' capacity to adapt to environmental risk and change in their natural and political-economic environment. However, we have never met an interpretation expressed in these terms by Nigerian smallholders, though evidence of specific components of the analysis (such as declining soil fertility, or the fall in prices of cotton or groundnuts) are not wanting in particular situations. The synthesis is, indeed, overtly challenged by another group of researchers who have found evidence of positive outcomes of migration (Mabogunje, 1972; Udo, 1975; Snrech *et al.*, 1994), superior income diversification strategies in the drier Sahel (Reardon *et al.*, 1992), innovative exploitation of new employment and market opportunities (Mortimore, 1989; Kimmage and Adams, 1992), containment of soil degradation under intensive farming (Mortimore, 1993a), and signs of a turn-around even in places where a linkage between income diversification (migration) and degradation has been routinely accepted by observers (Ford, 1997). The reader is already aware of the role accorded to flexible and adaptive choice, in a context of diversity, by the theoretical orientation of this study (Chapter 2). At this stage we need to take a closer look at the 'multiple stranding' of strategic choices by households and individuals.

Risk avoidance

It is by now a rather banal observation that income diversification is a means of spreading risk. Earlier work in the Nigerian region has shown that one practice in particular, the dry season circulation described by the Hausa term *cin rani*, performed such a function in exemplary manner, not only for the man who travelled (to the tree crop zone of south-western Nigeria, or the tin mines of the Jos Plateau), but for his household too, for whose granaries the burden of feeding was significantly lessened during his absence. This double benefit is reflected in the ambiguity of the English translations offered for the term: 'eating away the dry season' according to Prothero (1959, 1996). Such movements have equally deep roots elsewhere in the Sahel (notably in Burkina Faso: Rouch, 1956), and have established themselves as all but irreplaceable contributors to household economies across a swathe of southern Niger, also dependent on the markets of Nigeria (Rain, 1997).

A group of studies carried out during and shortly after the drought of 1972–1974 in northern Nigeria (summarised in Mortimore, 1982, 1989; Watts, 1983) and in Niger (Grégoire, 1980; Raynaut, 1980) allow some detail to be added to these observations. First, a variety of incomes is characteristic of households and also of most individuals, including women, who have a source of private income which they carefully maintain (known in Hausa as *sana'a*). Merely to illustrate their variety, we may cite as examples butchering, drumming, hair dressing, mat making, well digging and herbal medicine. Second, each has an economic strategy appropriate to his or her status and resources of capital, skill and social contacts. Third, wider economic choices are arranged on a ladder of rising unacceptability against the contingencies which are entailed by food shortages or other causes. For example, resistance to the divestment of assets, in order to enter the market for food, is relative to need and to personal choices available elsewhere. Resistance to selling personal and moveable assets (such as radios or cycles) is weaker than resistance to disposing of land. Fourth, resort to *cin rani* in a household, and to less acceptable choices, fluctuates from year to year inversely with the size of its grain reserves. The 'grain accounting year' begins at harvest time, when it is clear to every household head, if not to all its members, how many months' deficit they can expect. Fifth, in a major food shortage, the social incidence of *cin rani* and of less acceptable strategies broadens from those who participate regularly to those who are driven to it. Finally, as rural communities sought adjustment in the longer term to the lower rainfall conditions of the 1980s, the level of commitment to income sources outside the village deepened for many, if not all households. In the village of Dagaceri, those who first essayed the risky journey to Lagos in the 1970s, there

120

to sleep in the streets as night guards, later began to prosper as exporters of goats to the burgeoning festival meat market. This example links back to livestock production as another form of market-driven economic diversification for farmers.

Capitalisation

From the foregoing, it follows that it is as misleading to judge farming households' economic welfare solely in terms of their crop production as it is to treat them as dependent exclusively on the 'economic space' of the village territory. The intensifying regional integration of the Sahel with the coastal and sub-humid zones of West Africa during the past 30 years has been graphically demonstrated by the *West Africa Long Term Perspective Study* (Snrech *et al.*, 1994). But to what extent are these adjustments, made in myriad decisions at the household level and aggregating into regional trends, a result of risk avoidance rather than accumulative strategies, which were also given new dimensions and opportunities by the colonial interventions in West Africa? Markets have provided many such opportunities (Siddle and Swindell, 1990), amplifying the complexity of access to inputs, transport or intermediary profits on the part of persons privileged through their social status or role in government bureaucracies, and their access to land (Hill, 1972, 1977; Clough, 1984; Swindell and Iliya, 1992). For trading was, until recently, given a higher social value than farm production in most West African societies.

Our concern is with the management of natural resources, and an important question is therefore to what extent privately accumulated capital is recycled back into agriculture. To begin with, it is well known that wealth is invested in livestock, whether owned and managed by the same person, or contracted to livestock keepers by owners who may not even live in the countryside.

The implications of this deeply entrenched attitude to livestock investments for 'sustainable management' of the natural resources are ambivalent. The traditional view, based on 'carrying capacity' principles of range management, was that it encouraged the overgrazing and degradation of common pastures. But the value of livestock, in particular small ruminants, to the intensification process provides a major check to this view. Small animals are usually repositories of the savings of poor people, including both women and children, and so it is not surprising that livestock densities (even when small ruminants are converted into standard livestock units) correlate highly significantly with human population densities in northern Nigeria and Niger (Bourn and Wint, 1994), a correlation that was demonstrated in the Kano Close-Settled Zone by Hendy (1977). As this Zone contains the most intensive farming system found anywhere in northern Nigeria, and indications are that it is ecologically sustainable (Harris, 1996), it gives the lie to the claim that increasing densities of livestock necessarily lead to degradation (cf. several essays in Leach and Mearns, 1996b). It is astonishing that this evidence of a propensity to save – a precondition for

121

economic growth according to classical theory – was for so long denigrated by agricultural officials all over Africa as wholly negative.

A major area of farm investment, closely linked to livestock, is the use of animal energy (usually bulls) for plough and cart traction. Governments have operated credit schemes in the past to promote this as a part of 'mixed farming' packages, going back as far as 1927 in Nigeria (Buchanan and Pugh, 1955). In Dagaceri, all the seven plough teams in operation in 1972 (which gave 42 per cent of the farmers some access to their services) were lost by sale or mortality during the ensuing drought. Only 4 per cent used ploughs in 1974, borrowed from another village. Plough ownership did not reappear until 1979, when two farmers bought ploughs and bulls; in the following year there were four, and by 1985 there were seven. This twelve-year cycle of recapitalisation was not, however, financed from the proceeds of farming, as groundnut production had collapsed in 1975 and did not recover. One of the first two owners was a tailor, and all the new owners had obtained funds from seasonal migration. This experience of the investment of off-farm income in agriculture, which is not untypical in Nigeria, is apparently very different from those of villages in Mauritania and Niger where, according to studies carried out at the same time (Bradley *et al.*, 1977; Raynaut, 1980), agricultural investment had to compete with many other priorities and was very low.

We proceed now to the third and fourth steps of 'multiple stranding'. According to Haggblade *et al.* (1989), there is, in Africa generally, 'a rural structural transformation involving increasing specialisation and diversification out of agriculture'. It is possible to reconstruct the off-farm sector of household economies in a part of the Kano Close-Settled Zone using data from baseline and later surveys (Mortimore and Wilson, 1965; Amerena, 1982; Mortimore, 1993a).

Many households are not self-sufficient in food grain and have not been so for at least four decades, owing partly to a shift, early in colonial times, into groundnut production for the market and partly to an insufficiency of land-holdings after generations of subdivision. In 1964, when the baseline study was conducted in three villages within 15 km of Kano City, eight specialised and 20 less specialised secondary occupations generated income for 87 per cent of household heads. Average income amounted to about 13 per cent of the total estimated market value of their farm and non-farm production. This excluded earnings by migrants and by women. In 1979–1980, in two different villages at 7 and 23 km from Kano, non-farm occupations generated 80 and 47 per cent (respectively) of total household cash income. Only 22 per cent and 26 per cent of the households produced more than their minimum grain requirement (estimated to be 218 kg per capita). To meet this contingency, increasing amounts of labour had been diverted into non-farm occupations (Amerena, 1982).

During the ten years 1969–1970 to 1979–1980, non-farm occupations increased as the primary source of cash income from 57 to 90 per cent of households in the first village, and from 44 to 64 per cent in the second. Proximity to Kano City also correlated positively with expenditure on farming inputs and livestock ownership (except cattle). Amerena concluded that involvement in non-farm occupations increased with agricultural intensification and with access to the urban market, then experiencing the full effect of the Nigerian oil boom. But the major explanation of farmers' decisions to invest more time in income diversification lay in the catalytic effects of drought and groundnut disease. Some financed income diversification out of agricultural savings. But today, investment in ploughs and draft animals, fertilisation, hired labour and other inputs is often financed by off-farm incomes.

In this way, business and off-farm activities (categories 3 and 4) link back directly to the subject of Chapter 6. In Nigeria, unlike many countries (for example, in eastern and southern Africa), increasing integration with the market, and more especially the migrant labour market, does not necessarily lead to neglect of farming investment or to unsustainable management of soil resources.

Keeping animals

There is scarcely a smallholder farming system in the drylands of Africa that lacks an interest in livestock. But for decades, this duality in the goals and practices of rural households was ignored owing to the division of professional cadres between veterinary and agricultural departments in both colonial and post-colonial governments, which was mirrored in single-sector development research and interventions. Livestock breeders were portrayed in the literature as specialists having no farming interests, their mobile life-styles and cultural systems fascinating the researchers and their readers alike. The secondary livestock interests of farmers were largely ignored, as were the secondary farming interests of many livestock specialists. The consequence was a popular conception of two parallel but rarely intersecting production systems, a conception which remains firmly rooted even as research and development swiftly accommodate to a more accurate model.

At risk of oversimplification, the level of integration of livestock with crop production in dryland farming systems can be related to the level of farming intensity (Pingali et al., 1987; McIntire et al., 1992; Mortimore and Turner, 1993; Mortimore and Adams, in prep.). Put simply, the more frequently farmland is cultivated and sown to crops the greater is its need for enhanced fertility, and where alternative sources of nutrients are either absent or uneconomic, livestock manure is the preferred solution. As more land is transferred from natural rangeland to farmland, cultivated every year or nearly so, livestock come to depend more on crop residues and less on pastures for fodder. And as farmland becomes permanent, relatively free from bushes and roots, and farmers increase their cultivated areas to compensate for dropping yields per hectare, the use of animal energy for ploughing, weeding and farm transport becomes more attractive. Livestock are always valued as investments and therefore their numbers tend to increase with those of the human population, though there may be a switch of emphasis from cattle to small ruminants, which are more flexible to manage under smallholder intensification. Such a switch appears to have occurred in northern Nigeria (de Leeuw *et al.*, 1996).

Yet livestock production remains a distinctly different sector from crop production, no matter how integrated; its capitalisation, production objectives and mode of participation in markets are quite different. Thus, for a farming household engaged in gradual intensification, increasing its livestock interest is, at the same time, both diversifying its economic base and integrating its farming system. What then is the status of livestock husbandry in our four systems?

Kaska (Ibrahim, 1996)

Stock-owning patterns reflect ethnicity very strongly, but wealth and inclination are also important. Most Manga farming households keep sheep and goats in small numbers, and comparatively few own cattle (Table 7.1), which are managed by neighbouring FulBe in exchange for payments. Horses and donkeys are used for carrying people and loads respectively, while small ruminants are kept as assets.

All FulBe households, on the other hand, normally keep herds of cattle (on average between fifteen and 25 per household) as well as small ruminants (30 to 50 sheep or goats). Many FulBe own camels in addition to horses and donkeys.

Respondents reported that the population of livestock (particularly of cattle) was several times higher in 1960 or thereabouts, but has since fallen drastically and changed in structure: there are more goats than sheep, and fewer cattle, horses, donkeys and camels. These changes are said to be due to a scarcity of fodder and water. Furthermore, the ownership of livestock has changed. Around 1970, FulBe held about three-quarters of the livestock in the area, but now more animals are said to belong to Manga. They have a source of income from dry season *garka* cultivation, which has meant that in dry years they have not had to sell their livestock to the same extent as the FulBe in order to get food and fodder and to meet other social obligations.

Table 7.1 Households owning livestock in Kaska, 1992–1993

Livestock	Number of households	% Households
Goats	26	72
Sheep	26	72
Horses	10	28
Donkeys	7	19
Cattle	3	8

Source: Ibrahim (1996) (n=36).

Both Manga farmers and FulBe livestock producers graze their animals on the *tudu* grasslands and on abandoned or long fallowed *kwari* farms. Grazing land is open to Manga, FulBe or strangers' livestock; neither the FulBe nor the Manga prevent one another, or strangers, from grazing on available pasture, notwithstanding the rigid demarcation of village territories which governs access to land for cultivation. The FulBe are able to prevent the neighbouring Manga farmers from extending cultivation into their grazing lands, acquired through rights of prior settlement, if necessary by intentional crop destruction, verbal or even open combat.

During the wet season (June to September) FulBe livestock are confined to rangeland, but after harvest (October to early December) grazing is extended on to farmland, especially where free crop residues are available. Around March, if the fodder nearby diminishes, livestock may be moved to wetland areas until June. The sedentary Manga farmers, however, confine the movements of their small ruminants to within 2 or 3 km of the village. During the dry season, the small ruminants in the hamlet studied are managed by shepherds (or groups of shepherds) who are paid by each household on a monthly basis. In the wet season, however, when every adult is busy with farming, the animals from individual households are managed by boys and girls between the ages of 8 and 10 years. Although this system uses more *labourers*, it is more economical in its use of *farm labour*, as weeding is arduous work which demands as many adults as possible.

In these livestock breeding systems, there are some indicators of integration with crop production. Residues are grazed after harvest – but generally not restricted to farmers' own livestock; and manuring is by night corralling wherever livestock ownership and farming are combined in the same household, more especially among the FulBe, who maintain large millet farms in the *faya* under their control. However, the limits of integration are shown in such characteristics as residue wastage, inefficient channelling of manure from range to cultivated land, and an absence of ploughs and carts. The ethnic segregation of two rather different livestock systems allows ecological niches to be effectively exploited, but restricts the synergistic benefits. Of course, neither community keeps animals for integration. Capital accumulation and breeding for the market are their overriding objectives, on which recurring animal mortality persistently sets a ceiling. No one can either control or foretell the incidence of drought or disease. Everyone would own more animals if they could. In addition, FulBe women sell milk in weekly markets.

Futchimiram (Chiroma, 1996)

Although the Badowoi are crop cultivators, they also own livestock (camels, cattle, sheep and goats, horses and donkeys). Cattle are kept for breeding, milking and manuring, and by a few farmers for fattening and sale. It cannot be said that they are specialist livestock producers, yet they are more committed to livestock (and to cattle) than other farmers in northern Nigeria. Although migrant FulBe enter the area, the Badowoi neither like nor encourage them; they have no rights to land. The Badowoi assume rights to *all* land and take it into cultivation when they need it. Nothing is known (with any certainty) about trends in livestock ownership or numbers in Futchimiram.

Residues are grazed after harvest, the herds thereafter moving into the wooded parkland where they continue to graze during the rainy season. During the dry season, when fodder is low in quantity and quality, work animals (camels, donkeys and horses) and some small ruminants may be fed on cereal grains or the nutritious legume residues. Cattle are left with the options of grazing on crop residues, cereal bran, browse and available grasses. The feed resources may be low in protein and energy content, leading to loss of body weight and reduced milk production. Only lactating cows and old, sick and young cows receive priority in the provision of high energy supplements such as cereal grains, legume residues and cotton seed-cake. The nutritional stress often encountered by cattle also renders them more susceptible to opportunistic infections and even death if the dry season is prolonged. As the Badowoi never take their livestock outside the area, severe mortality may affect the herds after a major drought.

In the rainy season, between crop emergence and harvest, the cattle, sheep and goats are herded in the day time by children aged between 7 and 15 years, or by hired labour, in order to release adults for farm work (as in Kaska). At night they are penned beside the house for fear of crop damage. In the dry season, livestock range freely with minimal supervision, not because of labour shortage but owing to a lack of hazards. Not surprisingly, rustling has been reported.

Most Badowoi see livestock as investments, and as a form of security against contingencies such as low crop yields and social commitments (for example, weddings or naming ceremonies). Integrative indicators are residue grazing and the management of manure from the night pens, which is used on the home millet fields. Cattle owning for traction is not yet common in the area and carts are few (though both are now being promoted by the development agency). Badowoi do not produce milk for sale in markets. Donkeys, however (owing to their use in lifting water from wells), and horses or camels, are essential productive investments and are valued accordingly.

Dagaceri (Mohammed, 1996)

Here, as in Kaska, there is an ethnic basis for livestock specialisation. Cattle are mostly owned by the FulBe community, in herds of between ten and twenty animals, although a few Manga and immigrant Hausa households also have some cattle, sometimes managed by FulBe herders. Manga were more committed to cattle ownership in the past than they are now, according to tradition; but entry into plough and cart ownership during the past four decades has

Plate 7.1 Non-agricultural work: goats corralled in house compound, Dagaceri (with rain gauge).

brought several households into ownership of bulls. Both Manga and FulBe keep sheep and goats. Prior to the droughts of 1972–1974 and 1983–1984, many households – Manga or Hausa – in Dagaceri owned a horse, and many had donkeys. Currently very few households own either. Most FulBe households have a horse, if not also a donkey. However, the scale of FulBe livestock holdings is small, reflecting frequent subdivisions of inheritance and a major commitment to farming. Table 7.2 shows that livestock ownership is widely dispersed in the community, both among households and between individuals in those households. Of 20 owners of ruminants, thirteen are women.

Camel ownership has disappeared since 1970. Notwithstanding a visible pressure on land, it is not clear whether cattle numbers have fallen over the last few decades; while, in view of a growing human population, and trends observed elsewhere, an increase in the numbers of small ruminants is more likely than a decrease.

There are two systems of grazing management. Manga and Hausa farmers' sheep and goats are managed by someone appointed as the village herdsman. At the beginning of each day, children from each household take the animals to a rendezvous at the summons of a scout's whistle. Village flocks can attain a size of several hundred animals. In the dry season, the animals are taken to already harvested farmland (provided it has no *guna* melon planted in it), and to fallow land. During the rains, grazing is restricted to uncultivated rangeland and

Table 7.2 Livestock ownership in Dagaceri, 1994

Animal	No. of owning households	Mean per household	No. of owners	Mean per owner	Total
Twelve Manga households					
Sheep	9	7.4	15	4.5	67[1]
Goats	7	3.9	8	3.4	27
Plough bulls	3	2.3	3	2.3	7
Other cattle	1	4	1	4	4
Donkeys	2	1	2	1	2
One FulBe household					
Sheep	1	19	6	3.2	19
Cattle	1	15	5	3	15
Guinea-fowl	1	6	1	6	6
Hens	1	29	5	5.8	29

Source: data collected by Salisu Mohammed (July 1994).

Note
1 This includes nine rams.

fallows. At sundown, the flock returns and the animals disperse, obediently finding their way back to their owners' compounds. Bulls, and horses or donkeys (if any) are, however, fed on collected or purchased fodder at their owners' house throughout the year. The cost and labour of feeding is the major current investment involved in the transition to plough and cart farming. However, it is not regarded as a deterrent.

The FulBe living near Dagaceri move with their cattle, sheep and goats in search of pasture and water. In the farming season, they are restricted to reserved (open access) grazing areas and to cattle tracks and fallow fields found at the edges of the village lands. At the beginning of the dry season, they move to fallow and harvested farmland. Access to farms where *guna* is grown is only possible by negotiation, and only after it has been harvested. During a prolonged dry season, when fodder is scarce and stored feed exhausted, the animals graze on shrubby plants and lopped branches of trees such as *Faidherbia albida* and *Combretum glutinosum*. Transhumance – to the Hadejia River flood plain about 20 km south – was a general practice until the 1960s. It has since declined, coinciding with increasing difficulties of access to wetlands, and longer distance herd movements are not attempted as a rule. Many herds stay in the area throughout the year. Meanwhile, in anticipation of seasonal fodder shortages, in some years the Manga farmers cut and store grasses from their fallows before the FulBe herds can get to them, in order to sell the bundles later at prices the FulBe consider to be extortionate.

The main objectives of livestock ownership remain the accumulation of capital and a breeding capability, though under the more densely settled conditions of Dagaceri, large weekly markets where milk and butter may be sold are

more accessible. The Manga use a communal dry composting system for fertilising their inner fields. Pen droppings, straw and other household trash are
accumulated during the year on middens at the perimeter of the village, and
may be taken to the farms by anyone with the necessary transport. The FulBe
corral their animals on their own fields. Crop residues are increasingly privatised, and (as we have mentioned), so are fallow grasses when farmers so
choose. Cut-and-carry is increasing in importance with the greater numbers
of bulls, and as this must be of high quality, it is often carefully privatised earlier
in the year or purchased. There is thus some evidence of a trend towards greater
crop–livestock integration within the bounds of private farms. The exclusion of
FulBe herds from *guna* farms or from access to residues threatens however to
undermine the easygoing symbiosis of the past. Not surprisingly, there have
been disputes over stock damage to growing crops on private farmland.

Tumbau (Yusuf, 1996)

There is an ethnic basis for livestock ownership in Tumbau, but in quantitative
terms the FulBe interest has become very small as the Hausa farmers have taken
up more and more claims to the natural resources. Livestock play a very
important role in both the farming system and the economy. Most cattle are
owned and kept by sedentary FulBe, who have lived there for many years and
whose farming practice (unlike their counterparts' in Kaska and Dagaceri) is
indistinguishable from that of the Hausa majority. However, Hausa farmers
keep plough bulls. Otherwise, far more important to them are their small stock:
sheep and goats. A few households own donkeys for transporting manure, farm
produce, bricks and clay for building. Important personages own horses.
Migrant FulBe also visit the area during the dry season with mixed herds of
cattle and small ruminants.

Table 7.3 quantifies this livestock-producing system, as it works for our
twelve collaborating households. Four features deserve attention. First and
most remarkable of these is the sheer size of the livestock enterprise in an
acutely land-scarce crop-producing system. Taken together, converted into
standard tropical livestock units (TLUs), and spread evenly over the landholdings of these households, the animals represent 1 TLU per (cultivated) hectare
(46.1 TLUs on 45.7 ha). This is a valid indicator in Tumbau because there are
virtually no rangelands. A second notable feature is the wide dispersal of animal
ownership in the community. All twelve households own goats, and only one
has no sheep. No less than 40 individuals own goats, and 24 own sheep. These
include nearly all married women and a number of children; it is common for the
women and children in a household to own many more small ruminants than the
head. Bulls and donkeys, however, are always owned by household heads. A
third feature is the female bias in the sex structure of the small ruminant
population. The female:male ratios are: 4.8 for goats and 5.2 for sheep (notwithstanding the ceremonial value of rams for the Id-el-Kebir festival). This

Table 7.3 Livestock ownership in Tumbau, 1994

Animal	No. of owning households	Mean per household	No. of owners	Mean per owner	Total
Ewes	10	6.3	24	2.6	63
Rams	5	2.4	10	1.2	12
All sheep	11	6.8	24	3.1	75
Nanny goats	12	8.7	40	2.6	105
Billy goats	9	2.4	17	1.3	22
All goats	12	12.7	40	3.2	127
Plough bulls	4	2	4	2	8
Cows	1	2	1	2	2
Donkeys	4	1	4	1	4
Fowls	5[1]	32.8	not known	not known	164

Source: Data collected by Maharazu Yusuf (August 1994).

Note

1 Only five households enumerated.

provides the best indicator of market off-take, suggesting that a reproductive capability is a valuable source of household income (especially given the inflation of meat prices in urban markets in recent years). Small ruminants are capable of two conceptions a year, and twinning is common. A fourth feature (not shown in Table 7.3) is the reproductive success of these households which can be gauged from the fact that only 12 per cent of sheep and 9 per cent of goats were bought or received as gifts; the rest were born in the household.

Settled FulBe participate fully in rainfed farming, but some members of the family move south with the cattle (and some small ruminants) in the dry season in search of pasture. In general, the FulBe are finding it very difficult to graze their herds in the area; the only available spaces for them in the wet season are the cattle tracks, which are becoming narrower every year as farms encroach on them, and some strips along the major streams (the reaches of the Kano River in particular). Some nomadic FulBe herders pass through the area from Niger Republic in November, December and January and inflict a lot of damage on the farm trees by lopping branches without the consent of the owners, and this is becoming a source of conflict.

In the dry season the small ruminants owned by the Hausa range freely over the empty farmlands. Perennial, irrigated or long-season crops such as sugar, vegetables or cassava have to be fenced off. Most livestock management work (watering, tethering for the night) is done by women. Scarce stored feed (hay and straw residues) is kept to feed these stock until the first part of the rainy season, before the growth of farm weeds provides fresh fodder. The small ruminants are penned, by general agreement, in the first week of rainfall (May/June). Thereafter, grasses and weeds (collected from farms in the afternoon or early evening)

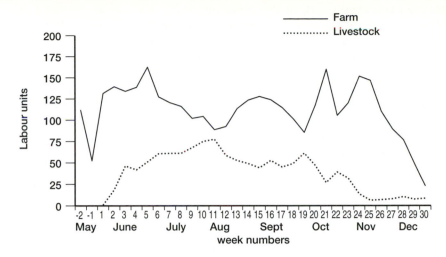

Figure 7.1 Labour use in farming and livestock work, Tumbau, 1996. (Labour use is shown in weighted units (equivalent to a man-day) per seven-day period.)

form their major feed until after the harvest of the last field crops late in November.

The maintainance of bulls and of horses or donkeys is a separate enterprise. They are fed throughout the year at their owners' compound, with stored or purchased fodder of good quality. Their increasing numbers (in recent years) pose an interesting challenge to an already intensive system: will it prove necessary to commit land to the production of fodder, or can enough residues, boundary grasses, stored hay or purchased imports be found? The need to provide for maximal energy output just at the time of the year when fodder is most scarce (in the early rains) mirrors exactly the dilemma of the savanna 'hungry season' for humans.

In Tumbau, time spent on farm and on livestock work during the growing season is negatively correlated, and the many animals consume a large amount of time. In the intervals between the weeding and harvesting labour peaks, an increased amount of time is spent on livestock work (Figure 7.1). Even in such an intensive, highly integrated system, not every household necessarily owns livestock. But to remain viable under the terms of such a system, the ownership of at least a few sheep or goats is essential.

Livestock work includes the management of animals during the day and at night, harvesting (or collecting) herbaceous fodder, and lopping browse from trees. In Tumbau, all animals must be fed in pens with gathered fodder throughout the cropping season, and each household therefore looks after its animals independently. Much of this is women's and children's work. Every day, fodder must be collected from harvested field weeds and field boundary plants (in particular, tall savanna grasses such as *Andropogon gayanus*, Hausa *gamba*), bundled and carried back to the compound where the animals are restricted during the cropping season, and their manure composted. Plough bulls require several times as much fodder as small ruminants, and need to be kept in the best possible condition if their energy potential is to be realised (not only on their owners' farms but on those of hire customers). A few horses, belonging only to wealthier persons, are similarly demanding. Donkeys must also be kept in good condition, as they are frequently hired out. Indeed, it is mainly their income-earning capabilities which justify the keeping of such time-consuming or expensive animals in households at Tumbau. Only fowls are free to scavenge at this time of the year, and even they need to be supplemented with chaff from threshing.

The indicators of crop–livestock integration are much in evidence: all fodders are assiduously collected, stored, privatised (and marketed); manure is composted as well as transported and distributed; traction animals are used for ploughing, weeding and transport; and livestock are managed within the task framework of farming households. Nevertheless, if asked why they own animals, few farmers will cite farm intensification as their objective: the value of animals as savings, and insurance against contingencies, is as prominent here as elsewhere. In addition, in Tumbau it is quite common for livestock to be bought for fattening and sold a few months later (for example, rams for festivals). A draft bull may be bought young, trained up and used on the farm for a few years, then sold for profit, using part of the proceeds to reinvest in another young animal. A farmer may travel to a distant cattle market to do this. At the same time, draft animals (like the donkey, which was ubiquitous before subsidised petrol undermined its competitive advantage) can be hired out, guaranteeing their owners income from ploughing during the early growing season, and from carting at other times. The fodder requirements of stable-fed animals are therefore becoming a matter for commercial calculation.

Conclusion

These four profiles of livestock management show how, under specific and differing sets of circumstances, households may assign strategic priorities to livestock. Critical in these decisions is the allocation of household labour between farming and livestock work during the growing season (during the dry season it can be assumed that demands on labour are less). The curves shown in Figure 7.1 represent aggregate solutions by individual households, each of them having assessed the trade-offs and opportunity costs involved. The importance of livestock in Tumbau households gives the lie to the idea that crop and animal production are incompatible; the lesson to be drawn from these profiles is that under smallholder conditions in the Sahel, *the more crops produced, the more livestock kept.* Yet this is more than a technical triumph for mixed farming. In terms of households' livelihood strategies, livestock represent primarily a form of income diversification, and secondarily a mode of intensification.

Trading, making and serving

In northern Nigeria, the non-agricultural sector of the rural economy has usually been taken for granted in planning and policy. This is partly because it fell under no professional arm of the government, and few data were ever collected on its operations. More important (probably) was a belief that it contributed little to incomes. This neglect contributed to a widespread perception of the rural economy as being confined to the production of primary goods (mostly for subsistence), and depending on towns for manufactured goods, services and trade. Yet in Nigeria, as elsewhere in the Sahel, there have long been buoyant systems of local and regional trading, both within rural areas and reaching out from them into towns and cities. Such trading systems have been best documented in Hausaland (Smith, 1955; Hill, 1972, 1977; Watts, 1983). Trade relations were not confined to high value or specialist commodities. A longstanding interaction between Kano City and its hinterland was expressed in the form of a large-scale traffic in fuelwood exchanged for city manure (Mortimore and Wilson, 1965; Mortimore, 1972; Cline-Cole *et al.*, 1990). Rural–urban interaction is at the heart of economic development in urban hinterlands between 1960 and 1990 throughout West Africa (Snrech *et al.*, 1994).

Northern Nigerian farmers engaged in production for export crop markets as soon as they could gain access to transport, particularly of cotton and groundnuts, and this provided the main theme of economic development until the oil boom changed the contours of the economic landscape in the 1970s (Helleiner, 1966). Thereafter, food production for urban markets began to usurp the export trade, for example, in the supply of irrigated vegetables and the related irrigated 'wheat boom' of the 1980s (Turner, 1997; Andrae and Beckman, 1985; Kimmage, 1991b; Kimmage and Adams, 1990, 1992). Rural areas could not remain unaffected by these developments. They drove a process of monetisation which reached even the smallest of villages. Parallel changes occurred across the border in Niger (Raynaut, 1970; Grégoire, 1978). It has been observed above that trading is usually perceived as the most effective route to private accumulation. However, the possibility that this sector might be as dynamic as the agricultural sector still goes against the grain of conventional wisdom, as it is assumed to depend on spin-off from the primary sectors, and there are few strategies available for boosting it as a developmental priority.

What are these business, manufacturing and service activities? Table 7.4 shows the major livelihood strategies cited by small samples of informants in the four villages in 1992–1993. These activities are 'secondary occupations' to farming which form an important part of individual livelihood strategies – described by the Hausa word, *sana'a*. Many other activities are pursued on a smaller scale or for shorter periods. In addition, owing to the small numbers interviewed, the list falls far short of a complete inventory of livelihood strategies used in each place.

Table 7.4 Non-agricultural livelihood strategies used in four villages, 1992–1993

Kaska (26 individuals)	Futchimiram (group)	Dagaceri (40 individuals)	Tumbau (14 household heads and others)
Arabic teaching		Arabic teaching	
Barbering		Barbering	Barbering
Building	Building	Begging[1]	Building
Butchery	Butchery		
Cooked food selling	Cooked food selling		
General retailing	General retailing		General retailing
Grain trading	Carpentry		Calabash repairing
Herbal medicine	Carting[1]	Carting, transporting[1]	
Kaba (palm) cutting		Carving	Carving
Kanwa (natron) trading[1]	Civil service	Civil service	
Kolanut selling	Cloth trading[1]	Cloth trading[1]	Cobbling
	Cobbling	Kolanut selling	Fodder trading
	Labouring[1]	Labouring[1]	Henna/*kuka* trading[1]
	Drumming		
Livestock trading[1]	Livestock trading[1]	Livestock trading[1]	Livestock trading[1]
Mat making	Groundnut oil making	Mat making	Mat making
Medicine selling	Contracting[1]		
Roasted meat selling	Quranic study[1]		Quranic study/malam
Schoolteaching		Schoolteaching	Sugar trading[1]
Sour milk selling		Weaving	
Tailoring	Tailoring	Wood selling[1]	
Watch/radio repairing	Thatching		Watch repairing
	Water selling		

Note
1 Activities normally carried on partly or wholly outside the village.

It is too easily assumed that 'dry season occupations' come to an end when farming starts, creating a seasonal dichotomy, almost a dual economy. The reality is far more complex. A need for income, or commitments already entered into, ensures that some activity continues alongside farming in some households, even through the peak of the farming period. We recorded 28 such activities in our regular labour monitoring. As our study was concerned with the management of labour during the farming period, rather than with the dry season economy, our discussion now focuses on how households resolve the counter-claims of farming, livestock and other work.

Food preparation and selling comprised most of those activities we designated as 'business'. One category is cooked or processed foods prepared by women and sold around the village, often by their daughters (such as bean cakes, unfermented millet gruel called *kunu*, and – where FulBe are concerned –

sour milk, called *nono*). Men cook and sell meat, which is otherwise sold raw, and morning or evening tea (though this is uncommon in villages). Snacks and stimulants (kolanuts, *aya* or tiger nuts, groundnuts) are also sold by children under parental supervision, and edibles such as mangoes and sugar cane when in season. Provisions such as salt, dry tea, sweets or washing powder are retailed from street tables by men, or their sons. These activities are entirely dependent on a local market, and the level of activity gives some idea of the amount of cash circulating among the ordinary folk.

Local raw materials, in particular the leaves of the immature dum palm or *kaba* (*Hyphaene thebaica*) which are freely available in the three drier villages, support manufacturing activities such as the making of mats, rope, baskets and cornstalk beds. Some are sold outside the villages via weekly markets which are visited by urban traders. Other activities we classified as manufacturing are well-digging, house building, watch repairing and sewing, which serve the local market.

Other services for the local market include a wide range of activities such as water carrying, barbering, nail cutting, provision of herbal medicines and many others which, along with those mentioned, provide a reminder that participation in the national or global market is not the only form of exchange in which rural people may be involved. These are, furthermore, long-established patterns.

Labour use in business activities (which includes all those mentioned above) during the farming season reflects quite stable commitments to earning income in these ways, whose importance to individuals must sometimes push farming into second place, at least insofar as other members of the household are available to cover. From Figure 7.2, which shows labour use in business in Dagaceri in 1995–1996, it can be seen that the variability in business work (though variable between years) was much less during the year, than it was in farming. The same is true in other villages and years except where the amount of labour committed is very small. In the dry season, of course, things are different. In three villages (though not in Dagaceri), business work escalated rapidly after the harvest, as well as short-term migration.

The levels of commitment to business activities in our four villages are compared in Figure 7.3 and Table 7.5.

The amount of time consumed by business, relative to farming, thus varied a great deal between villages, reaching remarkably high levels in Dagaceri, not-withstanding the fact that in this village, cultivated land per household is higher than elsewhere. Table 7.5 explores three possible explanatory variables for these differences (population density, number of households in the settlement, and distance to the nearest weekly market). The only one which appears influential is the number of households, which is an indicator of the power of the local market. Dagaceri, with 770 inhabitants (1996), can keep a number of micro-businesses going, and it is worth noting that in settlements where short-term migration occurs on a significant scale in the dry season, the local market for food, drink, provisions and services is largest, most hungry and thirsty during the farming season. In the small hamlets near Kaska and Futchimiram where our

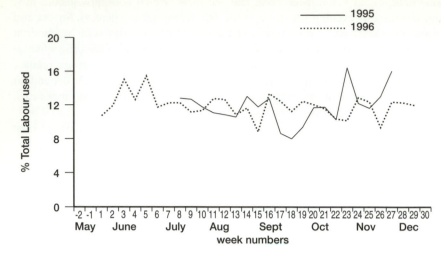

Figure 7.2 Labour use in business activities at Dagaceri, 1995–1996. (Percentages of total labour use are shown for each seven-day period.)

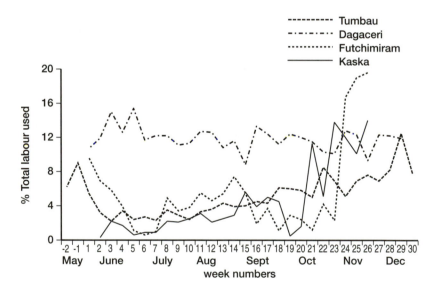

Figure 7.3 Labour use in business activities in four villages in 1996. (Percentages of total labour use are shown for each seven-day period.)

enquiries were carried out, almost no such facilities are on offer; people must walk to market for them on Fridays. But immediately the harvest of 1996 was over, the amount of labour given to business in Kaska multiplied by a factor of three.

Another reason for carrying on business activity during the farming season is

Table 7.5 Labour used in business in 1995 and 1996. Labour use is shown as a percentage of farming labour (average per seven-day period in the farming season)

Village	Population density/km^2	No. of households in settlement	Distance to weekly market (km)	Business labour 1995	1996
Tumbau	223	18	2	13	17
Dagaceri	43	157	13	43	51
Futchimiram	31	33	3	7	10
Kaska	11	36	5	6	10

an urgent need for cash at a time when food stores are often empty. Mat making in Dagaceri, for urban consumers to whom access is gained via local weekly markets, carries on throughout the year, and is still in evidence in July and August, when some men return from the fields to spend the afternoons working at it. It may be speculated that higher levels of commitment to business activity in 1996 were stimulated by a poor harvest in 1995, which in turn had followed a good one in 1994.

These are farming season patterns. However, a stereotype of rural house-holds' income diversification confined to dry season migration and restricted *in situ* to very low levels of activity, either by poverty or by the contrary demands of farming, requires revision – on the basis of these data – in the following three respects. First, the purchasing power of poor rural households, together with local markets collecting for urban outlets, can stimulate significant commit-ments of labour. Second, the farming season itself can, in some conditions, be a time of sustained business activity in the villages. Third, the level of labour–time commitment to business activity can, in some conditions, rise as high as 40–50 per cent of that devoted to farming, averaged over the season. Not surprising, therefore, is the finding that every adult in Dagaceri of either sex claims to have a craft or occupation capable of enhancing personal income. Thus do families attempt to broaden the basis of their livelihoods.

Labouring, schooling, marketing and travel

We have not yet exhausted the range of activities that compete with the family farm for labour time, even during the season when farming priorities are upper-most. There are four main categories of such activity: labouring (working on other people's farms); schooling (which competes for children's time); market-ing (trips to nearby rural periodic markets, or occasionally to cities); and travel to other places. It is the integration of all these subsectors of the household economy that necessitates our understanding them along with farming, animal husbandry and other income-earning activities. Villagers do not divide their day into 'economic' and 'social' sectors and, although most social activity takes place at night during the farming season, there are occasions (such as a funeral in a nearby village) when almost everyone (that is, the men and/or the women) may be called away for several hours. The four types of activity on which we focus now illustrate the difficulty of driving a wedge between 'productive' and 'unproductive' time, for in the mixture of social and economic, private and communal advantage we come up against the fact that the productive system is itself a social construct, reflecting the balance between contending interests and continually adapting to them (Berry, 1989).

The two distinguishing features of 'off-farm' activities as we have categorised them here are (1) that they make use of land and capital outside the household's own resources of fields, animals, skills and investments; and (2) for this reason, there is an interactive element in them by which they are embedded in the social matrix.

Labouring

Certain institutional arrangements permit farm labour to be redistributed among households. In purely economic terms, there is often a need to respond to a poor fit between the labour and land resources of the household. In social terms, labour sharing cemented communities – kinship-based or otherwise – in the past, as well as often making practical sense in particular farming operations. Under present conditions, labour-sharing institutions are apparently subject to rapid change. These three facets – economic rationality, social demands and changing circumstances – make for some complexity, as illustrated in detailed studies of the Kofyar farming system in Plateau State (Stone et al., 1990; Stone, 1997).

The labour exchange institution known by the Hausa word, gayya, long supposed to be in decline owing to pressures to privatise the economic benefits of farm labour, is still very much in evidence in all our villages, and a household's cooperation (or lack of it) still has a bearing on its status in the community. Nowadays the term has become flexible and covers differing arrangements, including teams of boys who are sent by their fathers to represent them, or are even recruited directly, for example, on behalf of the village malam (who is entitled to some help with farm work on account of religious services). The essence of gayya from a management perspective is that it concentrates labour for the rapid conclusion of a task, such as weeding, which will otherwise take many days. Insofar as it is reciprocated during the course of the season, it does not necessarily add to, or redistribute, the total stock of labour, but by rescheduling the territorial distribution of work, it breaks bottlenecks for individual households. The term gayya was employed loosely in our study to include neighbourly assistance, or help in an emergency (such as sickness).

Work on other people's farms may also be undertaken for hire. Most common is hand hoeing during weeding time, for there is a preference for keeping harvest work within the family. Such work (kwadago) is a long-established practice, and not something which began with currency; however, the export production booms of the present century have extended its frequency and acceptance. It is so common that labour hiring and labour selling have been taken as indicators of economic differentiation in the community (e.g. Hill, 1972). The institution tends to redistribute labour from poorer to wealthier households, from small to larger landholdings, and is often used by poor people as a short-term solution to food or cash scarcity during the farming season. When working for hire, of course, a man may not be able to give adequate time to his own crops, accentuating the difference in outcomes at harvest time.

Also in evidence, in Dagaceri and Tumbau, is hired work with ploughs. The 'driver' of a plough team may be employed by its owner to go around other people's farms for hired ridging or weeding work, analagous to an urban taxi. As this calls for uncommon skills, it is better paid. Thus the amount of time spent in working on others' farms – which in all our villages is usually within the community – is now rather an unreliable indicator of poverty.

Schooling

Primary school attendance in rural areas of the north of Nigeria has been marked by high rates of absenteeism, especially during the farming season, a reluctance of parents to register all their children, and official frustration expressed in high profile attempts to improve matters. Resistance to universal primary education (UPE), which was introduced by the Federal Government in 1976, focuses on two issues. First is a suspicion, commonly held among rural households, that formal 'western' or 'book' education in government primary schools undermines Islamic belief and practice, a suspicion that has not been fully allayed by the introduction of 'Arabic' or Quranic instruction by *malams* appointed to the staff of the schools. Second is the need for child labour on the farm. Teachers (who often leave the village at weekends) are sometimes seen as alien influences on village culture. Underfunding of school buildings, staff and equipment have made matters worse. These problems have arisen since the introduction of UPE in 1976 (Bray, 1981), prior to which, primary schools were few and selective. In 1976, it was expected that there would be an exodus of school leavers from the village and from the farm labour supply; but in fact most remain at home and return to farming, where the benefits of a formal education seem scant indeed. There is little evidence of any linkage between education and economic advancement in a country full of successful, innumerate traders. Hours spent by children in our villages at the government schools are therefore an outcome of quite a complicated tug-of-war between opposing forces.

More commitment is given in these villages to attendance at Quranic classes which are usually held in the village in the afternoons. Later, boys may be sent away to study elsewhere in the country under *malams* whose reputation has spread. These schools may take boys (*almajirai*) away for months at a time, even during the farming season, when their labour is actually used by their

Plate 7.2 Children's work: boy learning arabic, Tumbau.

malams on their own farms. For individual households, though not for the community as a whole, this can be a significant constraint; in Dagaceri, an asthmatic man who was quite unable to do heavy work lost access to all his three sons' labour in this way.

Marketing

People also travel outside the village, most importantly to markets. Weekly or more frequent market attendance is found, even at the height of the farming season; some individuals travel extensively for several days at a time, making arrangements for brothers or sons to carry on farm work in their absence. In economic terms, markets provide opportunities to buy inputs, durables and consumption goods, and to sell outputs. In the past, export crops were also brought to licensed buying agents whose abandoned scales can still be seen in some places. Prominent both visually and financially, therefore, are the grain and livestock dealers.

Owing to a preoccupation with neo-classical models of markets, the broader significance of rural periodic markets – and of informal, unregistered markets – is often not appreciated. In attending a market for several hours, a person may accomplish very little in terms of measurable economic transactions. As some people attend more than one market each week, and this even in the farming season, the priority given to market attendance in rural areas – although unquestioned by those who live there – is not easy to rationalise in economic terms.

However, markets should be understood in a much broader context. Social interaction occurs across a wider spectrum than that of the village, and market-days are in a sense holidays. When they occur on Fridays – the most common market-day in northern Nigeria – economic needs converge with patterns of religious observance. The market links rural people with the wider urbanised society through the visiting traders. Technological change is furthered at markets. Not only are they visited by both public and private sector promoters of new products and methods, but new seed varieties and technologies spread autonomously from market to market and across international frontiers (the *ashasha* hoe, and *Dan arba'in* millet, which both entered Nigeria from Niger, are cases in point). Input and output price information is updated at markets every week. This is not only important for determining farming and livestock strategies, business, manufacturing and other off-farm opportunities, but also plays a crucial role in decisions about the management of household grain stocks during the year. The market may also provide a forum for managing debt relations. Many markets coincide with sessions of legal or quasi-legal authorities where disputes are settled.

The rural market therefore functions as a node, linking the rural economy and society to the national system, and ultimately to the global economic system. Visits to markets should be conceived as 'management interludes' when household heads, and others, reach decisions about the optimal disposition of their resources. Having said this, however, it must be admitted that attendance at markets usually drops during the height of the farming season, when many matters are postponed, and peaks during the dry season.

Other travel

Dry season circulation, *cin rani*, is in suspension during the farming season. Nevertheless, not all travel ceases. A few examples may be cited. When the Id-el-Kebir festival falls during the rainy season, Dagaceri goat traders find ways and means of taking their consignments to Lagos, notwithstanding the demands of their farms. One farmer in Tumbau regularly leaves home for several weeks at the height of the growing season, having weeded his farms daily until his departure. Some Quranic scholars do not return home, even in the rains.

And married women, who under a patrilocal system need to visit relatives in neighbouring villages from time to time, do not necessarily stay at home throughout the farming season. Court cases, although infrequent, may call people away long distances.

Syntheses and trade-offs

A certain minimum of activity in labour sharing, schooling, marketing and travel is necessary to maintain a rural livelihood system which is a long way from being confined to 'subsistence' production. Figure 7.4 shows that, on the whole, labour commitments to these activities – lumped together – are relatively invariant through the farming season, which supports this inference. If anything, they tend to rise towards the end of the farming season. We may guess that reduced pressure of weeding work is not the only reason: dependence on purchased food reaches its highest level just before the millet harvest; and when harvesting starts, debts fall due, sales have to be arranged, and social business resumed.

However, between the villages, average levels of commitment vary a good deal. Figure 7.5 suggests a falling gradient from Kaska, through Dagaceri, to Futchimiram and Tumbau. Two possible explanations of this suggest themselves.

The first explanation concerns differences in market accessibility. Dagaceri is furthest from a weekly market (see Table 7.5), but motor transport is often

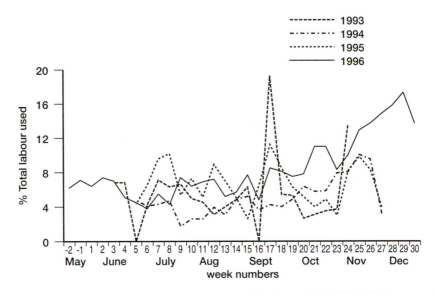

Figure 7.4 Labour use in 'off-farm activities' at Tumbau, 1993–1996. (Labour use is shown as percentages of total labour used in all tasks in each seven-day period.)

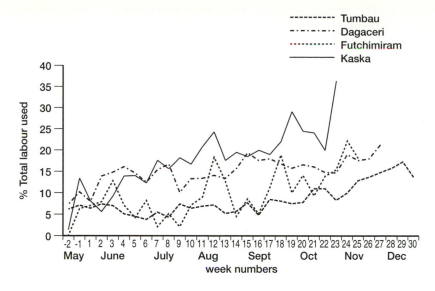

Figure 7.5 Labour use in 'off-farm activities' in four villages in 1996. (Labour use is shown as percentages of total labour used in all tasks in each seven-day period.)

available and some services are offered within the village. The study hamlet at Kaska, while nearer to a (very small) market, is separated from it by an arduous walk over sand-dunes. The study hamlets at Futchimiram and Tumbau are close to markets, but that at Tumbau is larger. The labour time spent in off-farm activities in 1996 reflected this grading quite exactly (Figure 7.5). This suggests a hypothesis that time spent in travel (a major component of off-farm activities), is negatively correlated with the accessibility of market services. This raises the 'real labour costs' of farming and of living in the ecologically and economically marginal places, a variable that is often cited but rarely measured.

An alternative hypothesis is that in the most intensive system (Tumbau), the demand for farm and livestock labour pushes out competing off-farm activities; in other words, the opportunity cost of farm labour is higher. It has been observed (Chapter 5) that the proportion of available labour used in farming is relatively low in Tumbau; however, when livestock work is taken into account this proportion rises to about 50 per cent at peak periods. Farm labour use in Kaska or Futchimiram rises, at times, to over 60 per cent. But this level of commitment is episodic, reflecting a small number of major crop management tasks that have to be rushed through in a short season, whereas in Tumbau, labour use is comparatively invariant, reflecting many tasks run in sequence during a longer season. In other words, the sheer toil of intensive farming leaves less room for off-farm activity, even though in peak periods labour is

143

under greater pressure in Kaska and Futchimiram owing to the more compressed farming season.

The high labour inputs to off-farm activity, including travel, in the three drier villages should be seen in the light of their historical profiles of engagement in the wider economy. In Dagaceri, early colonial trade in natron (obtained from seasonal ponds near Kaska, and delivered by camel to Kano and other markets further south) used to occupy many men during the dry season. In the 1950s, this was displaced by growing groundnuts for export, which required less travel, but when this crop failed in 1975, a new dry season migratory strategy, to exploit urban labour markets and the trade in goats to the south of Nigeria, established itself (Mortimore, 1989). In Kaska the natron trade goes back for centuries, and its control, indeed, formed the basis of the political economy of the region (Lovejoy, 1986). Natron (locally known as potash) is swept from the flat lake beds during the dry season, stacked, and exported by traders who send it through a distribution network to every corner of Nigeria. It is used as a salt cure for animals, medicine and a soup ingredient. There is still a small trade in refined table salt which, on account of its impurities, is tastier than the imported salt that all but swept it out of the Nigerian market. Although our collaborating households no longer make large commitments (as the *tafki* or natron pond nearest to them is no longer viable), this background has ensured a tradition of travel and interaction with the wider economy.

In Futchimiram, which is livestock breeding country *par excellence*, animals are marketed through Geidam, a major livestock market (56 km away) in the national system which is visited by trucks from as far away as south-east Nigeria. Although not many Badowoi from the Futchimiram area travel there frequently, some individuals may visit several local markets in a week.

Conclusion

During the farming season households work with a limited supply of family labour and, for most of them, access to additional labour from outside the household is severely constrained. Allocations of labour between farming tasks and competing demands are arrived at by several possible routes: authoritarian decisions of the household head, negotiations with individual members, independent individual initiatives, or recourse to social networks or financial resources (for outside labour). The way the labour force is used from day to day represents a balance of opportunity costs in which there are many potential trade-offs. Yet flexibility is severely constrained by the rainfall pattern and its effects on crop growth. For this reason, the role of non-farming work in the household labour economy is an essential part of the context in which natural and economic resources are managed to support livelihoods.

In this chapter we have pursued farming households, when they are not farming, through three steps of diversification (Figure 7.6). These steps are: first, livestock husbandry; second, business, manufacturing and services; and

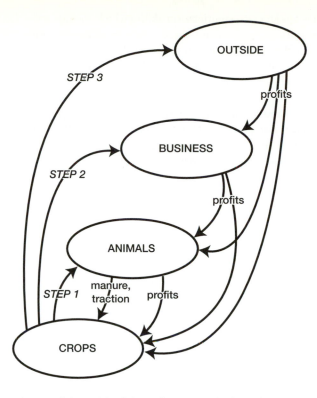

OUTSIDE

STEP 3

profits

BUSINESS

STEP 2

profits

ANIMALS

STEP 1 manure, traction profits

CROPS

Figure 7.6 A model of diversification in the household economy.

third, the 'off-farm' activities exemplified by labour sharing, schooling, marketing and travel. Although the categorisation is crude, it corresponds to significant
shifts in the relations between family labour and the other factors of production,
land, capital and skills.

In crop husbandry, all the factors of production are combined for primary
production, using the endowment of cultivated land to which the members of
the household enjoy rights of access. In the first step of diversification, livestock
husbandry, output from the secondary producers (animals) is added to that of
crops, access to additional natural resources (such as common rangeland)
becomes more important, and there are synergies in the form of manure and
animal traction which benefit the crop-producing enterprise. In the second step,
business, cultivated land becomes marginal, though common property resources
(such as wood, herbal medicines or water) may be important; however, capital
resources and skills mainly determine the options available. Profits from these
activities can be (and often are) returned to the crop-producing enterprise as
investments. Finally, in the third step, outside or other off-farm activities may
rely still less on the resources endowment of the household – either natural

resources or capital – to depend essentially on skills (which include knowledge, information and networks). Profits from these – where economic transactions are involved – can also be returned to the other levels.

The labour resource is divided among these levels of activity as an outcome of a continual process in which authority and negotiation are finely balanced in what Netting (1993) has called 'a network of implicit contracts' or a household. Clearly the anticipation and subsequent distribution of benefits are important considerations. The sheer complexity of the decision making itself – especially in larger households, which in Tumbau may exceed ten adults – suggests one reason why farmers are not always farming.

8

WOMEN, CHILDREN
AND THE HOUSE

Households, Islam and division of labour

In the past, it was often found convenient to adopt a view of the household as a coherent decision-making unit, a 'black box', with an integrated system of producing crops, livestock or incomes, and unified access to the resources needed to obtain them. Decisions of all the members were regarded as some-how summed in the household head, usually the person to be interviewed. The inadequacy of this view is obvious. Smallholder households do not consist only of standardised units of production and consumption, subjected to authoritarian decisions, but of individuals, structured by age and gender, whose decisions interact in a complex way. In this chapter, our objective is to move from the analysis of household labour use in the aggregate to a focus on the contributions made by women and children to securing household livelihoods, in particular in farming during the season.

An analysis of the division of labour on the basis of gender and age takes us into complex terrain, not only with regard to the collection and interpretation of data, but also because it necessarily enlarges the scope of the study to take account of cultural practice. This bland term hides some important subtleties. To begin with, there are divergences between accepted norms and actual practice which reflect the compromises every household must make between socially sanctioned behaviour and harsh economic necessity. We must then note that even if the norms are themselves invariable, there are differences from community to community in the extent of deviance that is considered tolerable. Finally, there are changes afoot, under the influence of pressures imposed either by endogenous driving forces (such as farming intensification) or by exogenous ones (such as social change).

The Islamic Jihad of Sokoto sought, from the first decade of the nineteenth century, to impose more rigorous standards of morality and practice on the nominally Muslim societies in the north of Nigeria, and to extend the Caliphate to non-Muslim areas. In cultural terms its impact was uneven, being greatest in the towns and least among non-Hausa ethnic groups in the countryside (Hiskett, 1984). The imposition of 'indirect rule' by the colonial government strengthened

147

the hands of the city-based Fulani rulers, who had used the Jihad to legitimise their appropriation of political authority in the emirates, but froze its expansion into the non-Muslim areas. Under the *Pax Brittanica*, the triple agency of the Hausa language (which became the lingua franca in much of the northern region of Nigeria), expanding markets and travelling traders and Muslim *malams* was instrumental in continuing to advance Islamic law and social practice.

However, the north-east of Nigeria, a vast region, subject to the authority of the Shehus of Bornu and their antecedents since the tenth century, resisted the jihadists. Islamic practice in this region had deeper historical roots, and the absence of a reformist ideology produced a more relaxed regime. The boundary between the Caliphate of Sokoto and the Sultanate of Bornu during the nineteenth century bestrode our four-village transect. Tumbau lies firmly inside the Emirate of Kano, and under Sokoto. Dagaceri and Kaska were under the rule of Machina, an outpost of Bornu, but during the past 30 years Dagaceri has accepted Hausa settlers from Damagaram (a region of the Sokoto Caliphate which fell under French colonial authority). As these settlers are a community of *almajirai* or self-confessed scholastic Muslims, this affects the practice of Islam in the village. Futchimiram lies deep in former Bornu.

This history does not affect land tenure, as Shari'a Law was recognised by the colonial government as the source of what was termed 'customary tenure' in all Muslim areas, and it has continued so to function. Shari'a Law does not prevent women from obtaining land – by inheritance, gift or purchase. The history does, however, affect gender participation in farming, as the practice of isolating married women (*kulle*), honoured as it is more in the breach than in the observance, is noticeably more visible in Tumbau and least so in Futchimiram. Dagaceri conforms more with Tumbau, and Kaska more with Futchimiram. This said, however, it must also be noted that both intensification and the influence of market forces are most pronounced in Tumbau, as it is within easy access of metropolitan Kano (50 km) and even more so of the huge, bustling, thrice-a-week market at Gezawa (15 km). Notwithstanding recent, unofficial, attempts to promote fundamentalist practice in Kano, intensifying contradictions between tradition and modernity are leading to a slow relaxation of restrictions on women's movements, even in urban areas.

Islam enjoins on married men the responsibility for feeding the household. The actual use of women's farm labour varies from household to household, from village to village and from task to task. There is much room for variable practice under differing conditions of labour scarcity, poverty and emergencies. The data show that women provide quite a small proportion of farm labour. Their contribution to domestic labour, livestock labour and some off-farm activities, however, is very large – the total amount of domestic work often exceeds farm work even during the growing season, when measured in standard units.

Deeply held values also affect children's participation in formal education

(Chapter 7). Village children who benefited from selective primary education before the introduction of UPE used to leave the area to seek urban employment, but few do so now. Fortunes, anyway, are perceived to be made as often through trade as by education; and the wealthiest men in all our four villages had little formal schooling.

Children from the collaborating households in Tumbau and Futchimiram must walk 2 or 3 km in order to attend a government primary school, and those in Kaska more than 5 km. Only in Dagaceri is there a school.

Children enter the labour force at age 5 or 6 or even earlier, and quite rapidly acquire the strength of an adult for farm work (by age 15 at the latest). Boys are highly valued on the farm, as they are strong and committed workers, and may even better their elders in mastering new technologies such as the plough. There is a dichotomy between those who opt to migrate (for education or employment) as soon as they are allowed to and those who opt to stay, who have everything to gain from hard work on land which will become their inheritance. Apart from in Futchimiram, girls are not asked to do heavy field-work as a rule, being most commonly employed in planting, harvesting or domestic work. Schooling is therefore not allowed to stand in the way of children's work. Quranic schools operate in the afternoons or evenings, when farm work is over for the day. But formal schools in some villages are virtually idle during the farming season, or families compromise with the imperatives frequently issued by government officials, by sending girls and retaining the boys.

Household demography

The persistence of high fertility strategies among Sahelian households reflects both high infant and child mortality levels and the value of children in the household livelihood portfolio. Children, especially males, provide additional farm labour and potential income diversifying agents. The demographic fortunes of households are therefore evaluated primarily in terms of success in producing and succouring children. The costs of having children are low where, from age 5 or 6 years, they can make a significant contribution to farm or domestic tasks, and from early adulthood they can travel for trading or labouring employment.

Demographic accidents such as mortality and infertility interfere with high fertility strategies of households, and largely account for the variable performances in achieving an adequate labour force. Within the social and economic context of a particular village, what matters is not merely the demographic size of the household but also the fit between labour and resources, most particularly farmland (for farmers) or livestock numbers (for stockowners). There is considerable variability both in the quality of fit and in the absolute sizes of households between and within villages (Table 8.1).

The adult labour force varies from an average of only 2.2 man-equivalents per

149

Table 8.1 Household sizes in four villages, 1996

	Tumbau	*Dagaceri*	*Futchimiram*	*Kaska*
Average household size				
Men	4.9	3.6	2.0	3.3
Women	5.7	4.7	2.1	3.1
Children	5.2	4.3	1.1	2.7
Whole household	10.6	8.3	4.1	7.2
Weighted labour units per household				
Average per household	4.7	3.2	2.2	4.2
Largest household	12.3	7.8	3.9	8.9
Smallest household	2.0	2.2	1.7	1.7

Note
Weights assign values of <1 to elderly men (Chapter 2).

household in Futchimiram to 4.7 in Tumbau. This range is a reminder that generalisation is inappropriate when applied to Sahelian agrarian organisation. The largest households (men, married sons and polygamous marriages) are, in fact, found in Tumbau (where farms are the smallest) and in Kaska (where they are the largest).

Household or family labour is subject to biophysical limits. The first and most obvious of these is set by available energy. Under pressure, farmers can work for up to ten hours a day at critical farming operations, but under such high temperatures and humidity in the growing season, average sustainable work loads, though unmeasured, must be lower than this. The second is set by age, which reduces the time and effectivenenss of aged persons' labour. There are abundant examples of men aged up to about 60 working a full day at farming tasks, but at greater ages there is a marked reduction in the number of hours worked. The third is set by incapacity, usually through illness, which can be expected to strike many people during the wet farming season (when *Anopheles* mosquitoes, water-borne parasites and other disease vectors are at their greatest intensity, and nutritional deficiencies most widespread). Small households, with few workers, are most at risk.

Women farmers on women's fields

The rules of divisible inheritance allow for a woman, for example, the wife of the deceased, to acquire land. However, the most common means of her acquiring it is by gift of her husband. Widows – who among the Manga and Badowoi often live alone in the village of their former husband, if they are too old to return to their parental household or to remarry – are settled with small pieces of land. Elderly widows are entitled to expect the assistance of younger male relatives in cultivating their fields. Occasionally a wealthy wife may acquire

Table 8.2 Women's fields in three villages

	Tumbau	*Dagaceri*	*Futchimiram*
Number of collaborating households	12	13	6
Number in which women hold fields	2	7	6
Number of women's fields	4	10	14
Average size in hectares	0.2	0.48[1]	0.43
Total hectares	0.8	4.8[1]	6.0
Women's fields as a percentage of land farmed by collaborating households	0.2	4	23
Land scarcity (percentage of productive land cultivated in 1981/1990)	99	62	24

Note

1 Includes five fields estimated.

In Kaska the fields are not formally divided between women and men.

land through purchase or loan. An entrepreneurial drive – to produce crops for sale – can be fulfilled by hiring young men's labour.

If women's access to farmland is primarily dependent on gift, it is predictable that a scarcity of cultivated land must constrain the amount that they control. When we examine the distribution of women's fields in three of the villages (Table 8.2), this prediction is confirmed in an inverse relationship between land scarcity (expressed in the form of the cultivated fraction) and the proportion of farmland under women's control.

However, other variables also affect the picture. The stricter observance of seclusion in Tumbau – women, at least in theory, should not need to undertake heavy fieldwork – means that with the exception of three land-endowed wives (one of whom works unashamedly throughout the farming season on planting, weeding and harvesting with her husband's approval), women are seen in the fields less often than in the other places; while in Futchimiram, women work so consistently in the fields that during the farming season the entire village is deserted for most of the day, apart from the elderly and one or two children.

In Tumbau and Dagaceri, women's small fields are farmed (or neglected) according to their own discretion or ability; but in Futchimiram, Badowoi custom is quite different. In marrying, a man expects to allocate a field of economic size to each wife. Thereafter there is a divergence in practice. In most households, while wives cultivate millet in their fields, their husbands concentrate their effort on their own fields, only assisting in time of need (such as clearing and planting, or during the peak period of weeding). But an alternative arrangement is to operate a family workforce which rotates the fields in turn. Using either model, the millet output from all the fields finds its way into the subsistence stocks of the household. The Badowoi do not intercrop as a matter of course. Groundnuts for market are grown in small, high density patches within the millet fields of either men or women. Women's superior access to land in this system strongly favours their participation in market production (Chiroma, 1996).

Women workers in men's fields

If women's farming on their own fields appears to be economically marginal in three of these systems (that is, excluding Futchimiram), the same is not true of the women's contributions to farm work as a whole, as Table 8.3 shows. In the table, women's work in *any* fields is included. Girls' labour is also included in the data.

The mean values show, with striking clarity, a rising gradient from Tumbau to Futchimiram in the percentage of labour contributed by women in each of the three major task categories – planting, weeding/thinning and harvesting – and therefore also in total farm labour. Our first conclusion, therefore, is that this variable correlates positively with women's access to land and negatively with scarcity of cultivated land, as presented in the previous section. In particular, the much higher values for Futchimiram women suggest a linkage with the fact that all and not merely some wives have their own fields; yet, as they contribute 52 per cent of planting labour, and nevertheless control only 23 per cent of cultivated land (Table 8.2), they must make a massive additional contribution to planting the men's fields.

The variable ratios (shown in italics in Table 8.3) indicate that the pattern of differentiation between the villages is not constant for all three farm tasks. The steepest gradient is found for planting, and the shallowest for harvesting work.

- In *planting*, Dagaceri women contribute twice as much labour as Tumbau women, Kaska women a similar percentage to Dagaceri women, but Futchimiram women contribute twice as much again as Dagaceri women. As planting work increases with the larger fields of the drier villages, and the need very often to repeat planting owing to poor germination of the seed, the available workforce must be stretched.
- In *weeding/thinning*, Dagaceri women contribute only half as much again as Tumbau women, and in both villages this is a time when many households withdraw women from the fields. But in Kaska and Futchimiram the extensive areas planted demand a continuing effort, at least until the pattern of the rain becomes established.
- In *harvesting*, the contribution of women rises much more slowly from Tumbau through Dagaceri to Kaska and Futchimiram. However, this is partly because in Tumbau, the contribution of women is greater at harvest time than in other tasks.

There is a great deal of variation from year to year, which depends on the rainfall and the condition of growth of the crops and weeds. We may generalise this variability by stating that in the highly intensive system of Tumbau, where

Table 8.3 Farm labour by women, three years.[1] Labour use by task is shown as percentages of all labour used on and off the farm

Village and year		Planting		Weeding/thinning		Harvesting	
Tumbau	1993	11.7		9.8		9.9	
	1995	11.3		11.5		19.0	
	1996	8.4		4.5		19.3	
	mean	10.5	*1.0*	8.6	*1.0*	16.1	*1.0*
Dagaceri	1993	38.2		30.4		28.2	
	1995	20.8		6.8		20.8	
	1996	20.8		4.2		8.2	
	mean	25.4	*2.4*	13.8	*1.6*	19.1	*1.2*
Kaska	1993	31.0		18.1		13.2	
	1995	nd		35.0		30.6	
	1996	26.4		38.0		24.8	
	mean	28.7	*2.7*	30.4	*3.5*	22.9	*1.4*
Futchimiram	1993	43.2		25.3		26.2	
	1995	nd		39.4		44.6	
	1996	60.3		37.6		29.3	
	mean	51.7	*4.9*	34.1	*4.0*	33.4	*2.1*

Notes
nd no data.
Ratios shown in italics.
1 1994 data are not included.

wives' segregation is most influential, women (of all ages) provide less than one-fifth of farm labour; in the fixed but more extensive system of Dagaceri, a proportion varying from less than one-tenth to more than one-third, depending on the circumstances of a particular year; in the extensive rainfed system of Kaska, usually between a quarter and one-third; and in the extensive but different system of Futchimiram, where women have the largest stake in farming, from over a quarter to over a half.

If the women of Futchimiram, among our four villages, are the most committed to farming work, the level of the women's contribution is everywhere impressive, more so when it is considered that the figures shown in Table 8.3 are for *weighted* labour (female adults were weighted as 0.7 man-equivalents and older girls as 0.5). This contribution includes the work of female children. In terms of hours worked, women's work is even more important.

Substantial though this contribution is, how large a proportion of *available* female labour does it represent? If we chart this parameter through a single farming season (1996), the four systems are found to be sharply differentiated (Figure 8.1). Notwithstanding the size of their contribution, the women of Tumbau and Dagaceri were scarcely stretched to make it. But in Kaska and Futchimiram, a veritable labour crisis in the weeding period pulled more than

Plate 8.1 Women's work: weeding millet fields at Dagaceri, July 1994.

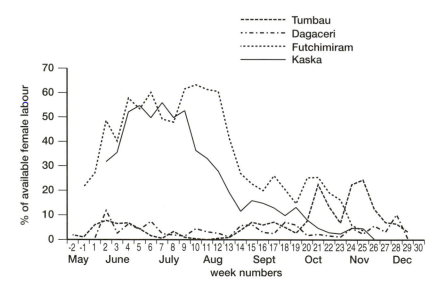

Figure 8.1 Percentage of available female labour used in farming, 1996.

half of available female labour out into the fields. The proportion of female labour used dropped abruptly in September when harvesting commenced.

We may conclude first by stating that women's farm labour provides a substantial resource without which the two driest and riskiest of our farming systems (those of Kaska and Futchimiram) would certainly not be viable, and the other two (those of Tumbau and Dagaceri) would most likely encounter unmanageable labour stresses at certain times. This leads us to draw attention, second, to the flexibility of women's labour, as it permits the total labour inputs which are achieved by men alone to be increased by a large ratio when necessary. Such flexibility is a crucially important asset in a Sahelian farming system. Finally, we emphasise the finding that, after putting aside the effects of rainfall, female farm labour inputs have an inverse relationship with land scarcity, and correlate positively with women's access to farmland.

It is important to note that our data relate to reported and observed labour use. They do not address the *control* of that labour, or the nature of negotiations between genders about work. Studies have begun to reveal the complexities of such intra-household negotiations in sub-Saharan Africa, and their significance for agriculture and other activities (e.g. Carney and Watts, 1990). Detailed qualitative research on intra-household decisions about the allocation of labour and its fruits would be of great interest, but lies beyond the scope of this study.

Child labour

We confine our attention here to the contribution made by children to farm work. This is not to deny the importance of girls' work (in particular) in helping with domestic tasks, fetching water or firewood, carrying food to the farm, and in retailing edibles under their mothers' supervision. Boys in their turn may sell their labour to other farmers. However, nearly all child labour is committed within the context of the household economy. Systematic exploitation of children in hired labour markets on the South Asian pattern is absent. It cannot be gainsaid that poverty is a factor increasing children's work load, as elsewhere (Johnson *et al.*, 1997), and that a labour shortage in the family is considered a good reason for having children, irrespective of their net value to the household (Cain, 1982).

Netting (1993) has described how smallholders depend on mobilising family labour for farm work, in a wide range of environments and historical epochs from old China to post-colonial Africa. Children are important to Sahelian smallholders, not only because they are the means whereby the family labour force reproduces itself for the future, but also because, from a very young age, they can themselves be a part of it. In our villages, children from the ages of 3 or 4 years may be seen in the fields at planting time. After their elders have prepared the seed holes, they can bury the seed with their bare toes as well as any adult. Even younger children may spend time in the fields on their mothers' backs or playing, while the family works at weeding or harvesting tasks. Resting and feeding under the shade of a farm tree, the family may spend the whole day outside the house, especially if the field is some distance from their village. From age 4 or 5 years upwards, healthy children can make a worthwhile contribution to fieldwork, and are taught farming skills. Miniature hoes are fashioned for them by the blacksmiths.

Boys take pride in working with gusto, especially when organised in peer groups, and by the age of 11 or 12 they may have the strength of an adult, though probably not the stamina – for weeding work, in particular, requires extended effort over many hours in high temperatures. Girls are not expected to work for as long as boys, or may be diverted to supporting tasks such as carrying cooked food from the house to the field. For young children, afternoon work in the fields is often excused, and they may attend Quranic classes in the village later on.

Random variability in fertility among men and their wives, and the risk of infant or child mortality, determine the growth or stagnation of the family labour force. However, there are other ways whereby children may be recruited to the house. Most important is looking after nephews or nieces from the houses of relatives. Another way is by taking on *almajirai* – boys aged from 8 to 12 – who can work on their *malam*'s farm. When an older householder, perhaps after a death or divorce, takes another wife, she may bring a stepchild with her. However, such arrangements are shortlived, for nieces are married off as young as 10 years (though 12–14 is more usual), and *almajirai* move on. The child labour resource is therefore fluid from year to year.

An assessment of the child labour resources available to our collaborating households strikingly reinforces their potential contribution to the farm enterprise (Table 8.4). We consider child labour as that provided by persons under the age of 14 (or marriage, if younger), weighted at 0.5 of a man-equivalent (ages 8–14) and 0.3 (ages 4–7). By comparison, adult females are 0.7 and an elderly person 0.5 (female) or 0.7 (male; see Chapter 2).

On the basis of our (admittedly arbitrary) weightings, children comprised from 25 to 33 per cent of available family labour in 1996. A second characteristic of available child labour is its variability between years, as the young arrived, grew older (passing from one weight category to the next), married and departed – or, tragically often, died. Finally we may note the range between households. In many, a small number of children was compensated for by having more adults, and so the scale of farming was not necessarily adversely affected. In others, in particular where elderly and infirm persons lived without adult sons or 'borrowed' children, a scarcity of labour could threaten the continuation of farming.

How much farm work do children do? We offer an answer to this question not, as before, with aggregated statistics, but with the living evidence from two households selected in Tumbau and Kaska (Table 8.5). The first of these has four working children. There is a grandson of the male household head, aged

Table 8.4 Child labour available in three villages, 1993–1996

	1993	*1994*	*1995*	*1996*
Tumbau				
Average weighted labour units per household	0.8	1.0	1.0	1.1
Largest household	1.5	2.6	2.6	1.4
Smallest household	0	0	0	0.3
Percentage of household labour force, 1996				25
Dagaceri				
Average weighted labour units per household	0.9	0.9	0.7	0.8
Largest household	2.0	2.3	1.8	1.8
Smallest household	0.6	0	0	0
Percentage of household labour force, 1996				33
Kaska				
Average weighted labour units per household	0.9	1.0	1.0	1.0
Largest household	2.4	2.6	2.6	3.0
Smallest household	0	0	0	0
Percentage of household labour force, 1996				30

13, and an *almajira*, aged 12, who work together and have been all but fully inducted to adult working practices, being able (as the table shows) to carry out a wide range of farm operations. The only difference observed from men's working patterns is that they are usually allowed to rest in the afternoons, after spending the morning in the fields. Then there are two grand-daughters, aged 11 and 9 years. Their contribution to farm or livestock work is confined to helping collect fodder when it is urgently needed; otherwise, they spend long hours in the manner so characteristic of village girls in Hausaland: strolling the streets, selling edibles to passers-by. There are seven younger children in this large household of 24 persons.

In Kaska, our household also has four working children. The elder two are both girls, one of them, at 12 years old, almost ready for marriage, and the other aged 9. Then there is a son of 7 years, who will assume responsibility as quickly as his strength allows him; and a small girl of only 5 years. In the absence of an older boy, the eldest girl worked a heavy load on the farm, both mornings and afternoons, seven days a week, during the early weeding period. After a respite (recorded as domestic work), she returned to the fields for harvesting at the end of the season. The two middle children are responsible for the animals. Every day and all day they supervise grazing on the grasslands near the village. If they fall sick, their elder sister is transferred from fieldwork to livestock tending. This family brought out all four children to help with planting at the beginning of the season, and again for harvesting at the end. Five years old is not considered too young for this work. However, there is one younger child in the house who stayed with its mother.

Table 8.5 The use of children's labour on two household farms in Tumbau and Kaska, 1994[1]

Age/sex	Tumbau				Kaska			
Weighting	13/m	12/m	11/f	9/f	12/f	9/f	7/f	5/f
	0.5	0.5	0.5	0.5	0.5	0.5	0.3	0.3
Task	Weighted labour units over period of 95/96 days							
Farm and livestock work								
Planting millet, sorghum					1.2	1.0	0.4	0.8
Planting late millet	0.6	0.3						
Planting cowpea, sorghum	0.6	1.2						
Planting benniseed	0.3							
Planting rice	0.5	0.7						
Transplanting	0.6	0.6						
Weeding grain mixtures	14.0	14.4			14.7	2.0		
Weeding rice	0.6	0.6						
Fertilising	0.3	0.3						
Collecting grass fodder	17.0	14.1	4.5	4.8				
Grazing animals					2.5	44.5	26.4	1.2
Harvesting cowpea	0.6	0.6			2.8			
Harvesting millet	0.9	0.9			5.5			
Harvesting groundnuts	0.3	0.3						
Total farm, livestock	36.3	34.0	4.5	4.8	26.7	47.5	26.8	2.0
Other work								
Food selling (groundnuts)			16.1					
Food selling (kolanuts)				26.7				
All work	36.3	34.0	20.6	31.5	26.7	47.5	26.8	2.0
Available units	48	48	48	48	47.5	47.5	28.5	28.5
Percentage used	76	71	43	66	56	100	94	7
School attendance	1.8							
Sickness				2.0		0.5	0.6	0.6

Note
1 The periods used are as follows: Tumbau, 96 days (1 July to 4 October); Kaska (where the first rain did not arrive until 21 July), 95 days (22 July to 6 November). At Kaska, one day is omitted for want of data.

Table 8.5 shows that healthy children old enough for farm work can give up to three-quarters of the time physically available to them; that in minding livestock, which is not so strenuous, quite small children can be committed for virtually every daylit hour; and that even where farm or livestock work are less demanding, girls and small boys who live in villages large enough to offer a market spend 40–70 per cent of their time offering goods for sale while they gossip or play. It is superfluous, perhaps, to point out that in being thus socialised into work from their infancy, children are far from being needlessly exploited; their parents work even harder.

Household work

The work considered in this section is women's work, though a small amount of men's labour is also covered by the term 'housework'. This (housework) is the biggest sub-category of domestic work, and includes grinding grain, and all cooking, cleaning, washing and in-house activity. It may also include the side-lines pursued in odd moments by women for occasional cash income, such as making grass *faifai* covers for calabashes, and unpaid services such as hair-dressing. The other categories included – which are conducted outside the house and may require substantial blocks of time – are fetching water, threshing and carrying food to the farm (time spent fetching wood is kept to a minimum in the farming season). The results are indicative, if still too coarse in texture to reveal the complex detail of women's activities.

There are four reasons why domestic work may have been ignored in economic studies of rural households. First is the elementary consideration that it is inconvenient to measure, especially where social convention denies male researchers direct access to women. Second is the idea that it has only secondary importance: a (male?) view of inert time which can be raided with impunity by more pressing demands for farm or other 'productive' work. Third is a traditional expert view that it has a zero economic value, or at least that its value cannot be estimated in any terms easily compatible with crop or livestock production. Finally, and following on from this, it can have little or no significance for economic development (offering few opportunities for technical advancement), except where it obstructs work of higher value. A broader conception of development, which takes social as well as economic welfare into consideration, has come to challenge such a view, but tends to treat domestic work as a welfare issue only concerned with women.

Putting numbers to the time spent in domestic work is, indeed, full of difficulties. First among the characteristics of domestic work is the fact that it can absorb a very large amount of time. Such quantities must be interpreted with caution, however, in view of its second characteristic, which is that the diversity of tasks is great, and the third, namely that the intensity with which domestic tasks are pursued is variable, not only among tasks of a different nature, but according to the pressures of time and competing demands, personality and even mood. Such elasticity appears less in evidence in farm tasks or herding animals, which are relatively homogeneous and predictable. The question is whether a time comparison between, say, farm work and preparing food is meaningful. A fourth characteristic of much domestic work is that its scheduling is flexible; that is, many tasks are not time dependent and can be put off if it is necessary to go to the farm. A fifth is its valuation in cultural as

159

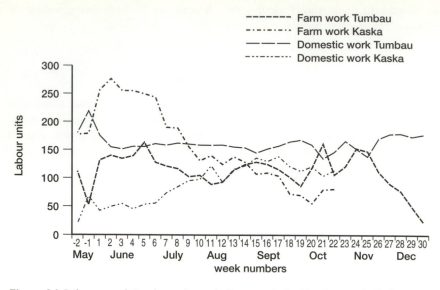

Figure 8.2 Labour used in domestic and farm work in Tumbau and Kaska, 1996. (Labour used is shown in weighted units (equivalent to a man-day) per seven-day period.)

much as (or more than) economic terms. It plays a central role in socialising children, especially girls, perhaps as a counterpart to 'play' in western culture.

Despite these qualifications, we have attempted to quantify domestic work in order to present a comprehensive picture of labour management in the household. Some domestic tasks are compelling, and therefore constrain the amount of 'available' labour that can be mobilised for farm or livestock work. As some domestic work centres on meeting the needs of children, there is a trade-off between reproducing the labour force of the future and meeting the demands of the moment. In effect, domestic work expands to fill every hour not appropriated by the other demands we have discussed above; leisure is a concept for the dry season only.

On the basis of what has been written already about women's participation in farming, and the data given in Figure 8.1, we may predict that domestic work commitments will be found to differ between Tumbau and Dagaceri on the one hand, and Kaska and Futchimiram on the other. Such a divergence is confirmed in Figure 8.2, which shows the amount of labour spent on domestic and farm work in Tumbau and Kaska in 1996.

In the 'Tumbau pattern' (which also holds for Dagaceri), domestic work occupies strikingly large quantities of weighted time, almost always greater than those given to farm work. A second characteristic of this pattern is its relatively invariant profile through the farming season. It does not matter that domestic work, as has been emphasised, includes a variety of activities, some of

160

them hardly comparable in either effort or economic value with those of farm labour. The point which has some interest is that the peaks in the demand for farm labour – weeding and harvesting – are clearly not being met, to any major extent, by diverting women from domestic work.

On the other hand, the 'Kaska pattern' (which also holds for Futchimiram) vividly shows a suppression of domestic work during June and July in favour of farm work – which is, of course, predominantly weeding work. The curve of domestic labour rises steadily, from a very low level, throughout this period, eventually passing that for farm labour in September. Here is a system where everything at home is sacrificed for productivity in the fields. Women's work is critical.

The shorter the season, the more concentrated and episodic the peak demand periods for labour become, and the greater the need to mobilise women as well as men. While individual operations may be just as urgent in Tumbau, in aggregate, levels of demand are lower and can be handled by skilful management of a continuous, largely male, labour force.

Sickness

For men as well as women, the rainy season brings increased morbidity and mortality to Sahelian villages. Combined with the effects of undernutrition and hard work, exposure to disease vectors benefiting from the higher temperatures and humidity is only partly compensated by additional vitamins and prophylactic substances contained in newly available vegetable food sources at this time (Etkin and Ross, 1982). Indicators of these sources of stress are the categories 'resting' and 'sickness', which were included in the labour monitoring exercise (Figure 8.3).

At Tumbau during the period from July to November, resting consumed, on average, 25 per cent of the time given to farm and livestock work together. An absence of sharp fluctuations in these variables taken together (farm and livestock work tend to compensate for one another) reinforces our earlier conclusion that this is a relatively invariant system. However, sickness, as we would expect, behaved differently. Very little was recorded until the end of August, when the number of man-days lost jumped, in successive weeks, from six to 27. During the first week in September, sickness consumed 20 per cent as much time as farm and livestock work together, and 38 per cent as much time as farm work alone. Of course, much of this impact was absorbed by a few households. The data on farm labour show that this crisis coincided with the all-important switch from weeding work to harvesting early millet. Had it occurred earlier, or later, its economic impact would have been greater. Thereafter sickness continued to consume a small but significant amount of labour time until mid-November.

Kaska, with its fluctuating, episodic patterns of labour use, makes a contrast with Tumbau. Here, as we have already observed, there is little opportunity for

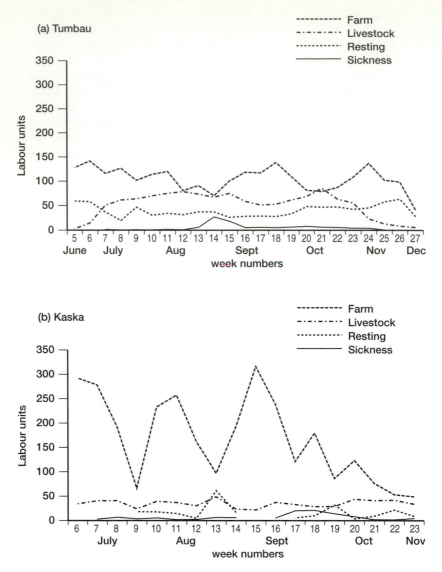

Figure 8.3 Labour time spent in resting or sickness in Tumbau and Kaska, 1995. (Time is shown in weighted units (equivalent to a man-day) per seven-day period.)

resting; on average, this use of time accounted for only 7.7 per cent of the time given to farm and livestock work between July and October. There was, however, a blip late in August, when tired families took a break between the second weeding and the harvesting of millet. Sickness did not assume importance until later, when an outbreak lasting for several weeks peaked at 21 days late in

Plate 8.2 Resting: farmers relax following the afternoon prayers at Tumbau, September 1994.

September, which amounted to only 10 per cent of the time given to farm and livestock work during that week, and 12 per cent of the time given to farm work alone.

An attempt to estimate the economic cost of sickness in these conditions would be made difficult by its episodic occurrence in time and random impact on households; but these data show how it fits into a scenario of risk for Sahelian families. What is even more difficult to estimate is the effect on productive performance of the energy constraint imposed through age, physique and poor health.

Negotiated livelihoods

In this chapter, we have used a method of quantifying labour time to analyse the role of women's and children's labour in securing household livelihoods from farming and other work. We hope that we have exposed the myopia of looking at smallholders' production systems in terms of only one sector, or only one sex, or a limited range of solutions. The inclusion of women and children in the analysis increases still further the complexity and diversity of the subject.

How women's labour, and that of children, is mobilised in the context of the household's full range of interests indicates a diversity of solutions, which are affected by the rainfall (its amount, variability and peak distribution), by scarcity of cultivated land, by women's access to land, by the practice of seclusion and by customary division of labour, as well as by the evolving dynamics of intra-household relations (not least gender relations). This diversity is fully consistent with the findings of a study of women's and children's roles in relation to male migration in the Sahel (David *et al.*, 1995), which cautions against an oversimplified scenario of inequity, deteriorating access to, and management of, natural resources, and increasing work loads for women, a scenario which has acquired a flavour of orthodoxy in some quarters (e.g. World Bank, 1995).

These numbers should be understood as the outcome of a process of negotiation that goes on continuously, as the uncertainties of the rainfall, the macroeconomic environment, and other factors superimpose themselves on the more predictable pattern of the seasons. In this process, the internal structure of the co-residential household is important. It will no longer do to regard it as an authoritarian hierarchy subject to the exclusive control of the (male) head. Such a model was always inaccurate (though many decisions *are* made in this way). Neither is it accurate to disaggregate the household into its individual members and to pit women against men in a competitive struggle. That the household may arrive at its decisions about the use of labour (its major and most flexible resource) in a complex and varying way is implied in the diversity of patterns we have uncovered, in only four small communities. What is also implicit is the functional effectiveness of the negotiations that take place. These smallholder-based farming and livelihood systems are demonstrably viable and resilient, notwithstanding claims to the contrary. At the heart of their resilience lies a powerful ethical and social commitment to work which – apart from infants and the infirm – embraces everyone in the family livelihood enterprise.

9

UNDERSTANDING INEQUALITY

Introduction

Debates about economic differentiation in rural African communities tend to centre on processes of distribution (or maldistribution) of wealth which are linked with political-economic changes in wider society. These changes have included historical interventions such as the arrival of the colonial powers, and their selective use of local elites to establish and further their regimes; or the promotion of export crop production, and the associated fortunes that were waiting to be made from cornering the profit margins on groundnuts or cotton; or (in Nigeria) the fall-out of the oil boom and its enormous system of patronage, whose tentacles spread from the Federal Government into every local government area of the country. There was also a vigorous debate during the 1980s, in which drought and famine were represented as destructive forces on the social order ('moral economy') which, to a large extent, was believed to have earlier protected rural households from the irreversible effects of asset loss and incapacity to feed themselves (Watts, 1984). The importance of taking account of such exogenous forces is evident, though empirical data to demonstrate worsening inequalities is not always easy to find, or to evaluate objectively.

The competition which differentiates rural society does not begin on a level playing field. At the local level, households are already unequally endowed with labour and natural resources, capital and management capabilities. Such inequalities are as old as society itself and, subject to the baleful influence of the global forces just exemplified, households struggle to make good their deficiencies or to exploit their advantages. The meshing of local with global forces affecting the distribution of wealth in society will not be understood unless they are confronted empirically at the local scale.

In this chapter we present six case studies of households (Table 9.1), selected from all four villages, in order to illustrate some of the ways in which smallholders respond to the circumstances in which they find themselves.[1] Beginning with their endowments of labour and land, and how these change from year to year, we proceed to brief descriptions of how they manage their land and labour resources, and how they dispose of the latter among agricultural and other

165

Table 9.1 Six Sahelian households in 1995

	Persons	Labour units	Fields	Hectares cultivable	Ha/labour unit	Livestock units[1]	Income sources
1 Inuwa Tumbau	4	2.0	5	2.0	0.9	2.3	Few
2 Mikaila Tumbau	19	11.2	15	11.2	1.0	8.7	Crop and animal sales, *cin rani*
3 Male Dagaceri	3	2.2	3	5.6	2.5	0	Mats, his son
4 Salisu Dagaceri	10	6.7	1	14.6	2.2	19.8	Crop and animal sales
5 Manami Futchim	4.8	2.9	3	5.0	1.7	nd	Crop and animal sales
6 Madu Kaska	12	8.9	3	26.7	3.0	nd	Irrigation, *cin rani*

Notes
1 Excluding fowls.
nd No data.

activities from week to week during a representative year (1995); and we end with identifying six key variables in a simple model for describing poverty (or wealth) in rural Sahelian households.

Inuwa of Tumbau

Inuwa of Tumbau has a small household consisting only of himself, his wife, a small son aged 5 and an even smaller granddaughter (by a previous marriage), aged 4. Neither he (about 59) nor his wife (49) are young. The small size of this labour force (2.0 weighted units) requires them both to undertake field-work, notwithstanding custom which prefers married women to stay indoors during the day. His hope is that when the boy grows older, Inuwa will be in a stronger position, though he himself will be elderly.

Relatively well endowed with farmland, Inuwa had 1.5 ha in three small fields in the year 1993 (Figure 9.1). Only one of these (of 0.5 ha) had he inherited from his father, and this is situated about half a kilometre from his house in the village; the others (of 0.6 and 0.4 ha) he purchased, conveniently close to his first field. In addition to these fields, his wife had two of her own (each of 0.2 ha): the first she inherited and the second was purchased in 1995, increasing the family's cultivated land to almost 2 hectares. The household had a land: labour ratio of 0.9 cultivated ha/labour unit in 1996 (close to the average for the village). They keep fourteen small ruminants and a few fowls on this land.

The soils of Inuwa's farm are reddish-brown sandy loams. Although there may be changes in colour and in drainage properties from field to field (and

166

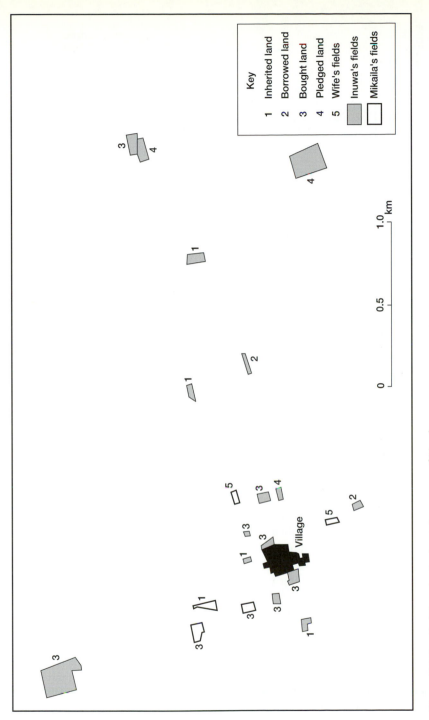

Figure 9.1 Landholdings in the intensive system of Tumbau.

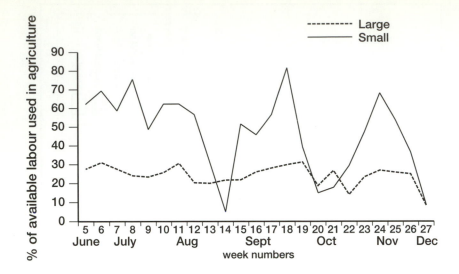

Figure 9.2 Labour used in farm work by a small (Inuwa) and a large (Mikaila) household in Tumbau. (Labour use is shown as percentages of available labour.)

even within a field, however small), the soils are generally low in fertility, free draining and easy to work. Relief is very subdued. He maintains these soils under annual cultivation by applying 4–5 tonnes/ha of farmyard manure, and, when affordable, small inorganic fertiliser inputs (70–90 kg/ha). Manure, owing to its scarcity, is rotated: thus in 1993, Inuwa manured two of his own fields, and in 1994 he manured the third one. In most years, early and late millet, sorghum, groundnut and cowpea are grown in fourfold mixtures, though the composition of the mixtures varies from year to year in each field, and groundnuts may be planted alone in a section of the field. Inuwa's own fields yielded 5–9 tonnes/ha of crops and fodder (residues and weeds) in 1993 and 1994 respectively (Harris and Bache, 1995). His wife also worked in her own fields and controlled their output.

In each year, Inuwa's household mobilised between 50 and 70 per cent of its available labour time – he and his wife, morning and afternoon, seven days a week, week after unrelenting week – for weeding the farms in June, July and August (see Figure 9.2). During these months plants always grow prolifically, provided that rainfall is sufficient, and the competition from weeds constantly threatens to inhibit the growth and yields of the grains and legumes. Then weeding slackened momentarily, before the labour inputs peaked again during the millet harvest. At this point, which in most years occurs from the first to the fourth week of September (depending on when planting took place), the household used over 70 per cent of its available labour time. Indeed, in 1993 they exceeded 100 per cent by hiring assistance. After another short rest of about

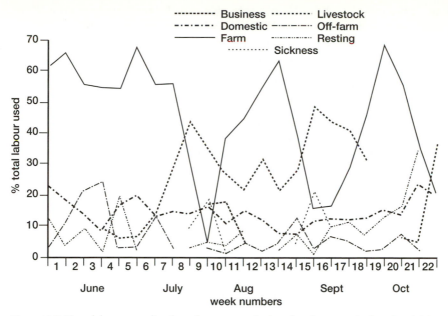

Figure 9.3 How labour was distributed among agricultural and non-agricultural activities by a small household (Inuwa) in Tumbau, 1995. (Labour used is shown as percentages of total labour use.)

two weeks, when use rates fell to 20–40 per cent, a second peak took them back to 60–70 per cent for the harvest of sorghum and late millet in October or November.

Notwithstanding the pressure of farm work, a large proportion of the labour used was devoted to livestock (Figure 9.3). In the year 1995, the three peaks which corresponded to weeding, millet harvesting and sorghum harvesting were separated by pronounced lulls in farm work. It was during these lulls that Inuwa took the opportunity to attend to a backlog of livestock work – mainly gathering and storing fodder. The short breaks also coincided with two short periods of sickness, early in August and late in September. It is important to stress the implications of such a commitment to work on the part of what is, after all, a *domestic* as well as a *producing* unit. It is not surprising that a low priority was given to domestic work, and still lower to business or off-farm activity; and resting, which had been an unaffordable luxury since May, rose suddenly, as if with a sigh of relief, after the sorghum harvest in November.

This pattern represents a high level of efficiency in mobilising the key resource in agriculture – family labour. With a small labour force on a small-holding, and with very low capital resources, Inuwa nevertheless achieved his family's subsistence needs, according to his own assessment of the harvests of 1992 and 1993, and according to estimates of nutritional requirements in 1993

169

and 1994 (Harris, 1996). He was even able to sell three bags of grain in 1992, and aimed to do the same in 1993. His wife's output was additional.

Though exemplary, this performance was by no means exceptional. Without additional labour – or labour-saving technologies, such as a transport animal or ox-plough – Inuwa cannot easily lessen his work load or increase the size of his farm (unless he reduces the intensity of its use). We are thus confronted with a paradox: in our most densely populated and intensively farmed system, labour *shortage* rather than *surplus* constrains real income. Farmers in Tumbau are keen to adopt ox-ploughing – notwithstanding their small and fragmented holdings – in order to break this constraint, but this transition requires substantial capital, either for buying the plough and team or, no less significantly for poor households such as Inuwa's, for hiring.

With such a smallholding, savings are difficult to accumulate from agriculture alone, though Inuwa sells groundnuts and cowpeas to finance his operations during the year. Larger contingencies are met by selling livestock. He sells dried henna or baobab leaves during the dry season, or makes a few mats with materials bought from traders coming from further north; such activities turn idleness into income, but do not generate savings. A small household, or an elderly person, is at a natural disadvantage with regard to *cin rani*, for the only person who could pursue alternative sources of income outside the village is Inuwa himself – at the cost of leaving his family for several months each year.

Mikaila of Tumbau

Mikaila of Tumbau could not be more different from Inuwa. The large household over which he presides (at the age of 63) contained seventeen residents in 1993 and had grown to nineteen by 1996. He has three married sons (aged 37, 30 and 25 years), two of whom, like himself, have two wives. Two grandsons, aged 15 and 12, are old enough to make substantial contributions to farm work (though one left for study after 1993), and as if this were not enough, his labour force is swelled from time to time by *almajirai*, one of whom was available in 1994 and 1995, and another in 1996. It is not surprising to find that there were four small boys and three girls promising yet more workers in the future, subject always to the risk introduced by a high level of infant and child mortality (three were born and two died during the four-year period). The household's labour force, furthermore, grew from 9.7 weighted units in 1993 to 11.2 in 1996, in striking contrast to Inuwa's small family.

With the married sons in residence, this household exemplifies the extended family production unit known in Hausaland as *gandu*. By agreement, sons continue to work on their father's farm until his death, while generating some income for themselves on fields which they acquire by his gift or through purchase or loan. This is the alternative to dividing the holding, and setting up sons on independent farms during the father's lifetime. Although once understood to be approaching obsolescence, under contemporary conditions

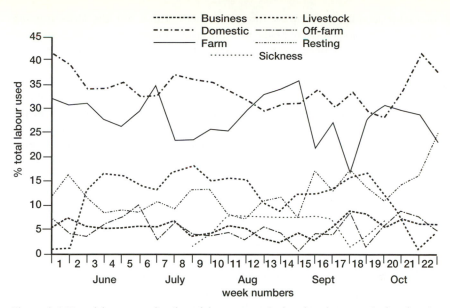

Figure 9.4 How labour was distributed between agricultural and non-agricultural activities by a large household (Mikaila) in Tumbau, 1995. (Labour use is shown as percentages of total labour used.)

of market production and privatised incomes (Goddard, 1969, 1973; Buntjer, 1970), the *gandu* has turned out to be more resilient than expected, whether because of land scarcity in such areas as the Kano Close-Settled Zone (Hill, 1972, 1977) or for other reasons.

Mikaila inherited only three of the thirteen fields he was using in the year 1993 (Figure 9.1). These were each fragments of larger fields belonging to his father which were subdivided between three or four sons on his death. They added up to less than a hectare (0.6, 0.3 and 0.1 ha). By 1993, he had obtained five additional fields by purchase (one of them as little as 0.1 ha, three more of less than 0.5 ha, and one exceptionally large field of 3.7 ha). These represent a very substantial investment in land whose market value is high and rising. However, these did not exhaust the acquisitive energies of Mikaila. The institution of *jingina* ('pledging' or taking land in exchange for a loan, redeemable on repayment) was used to gain access to three more fields of 2.9, 0.6 and 0.2 ha. Two more small fields of 0.3 and 0.2 ha were hired or 'borrowed' (*aro*). During the course of our study Mikaila was still acquiring land, hiring two more plots of 0.7 and 0.3 ha in 1994. His total holding in 1996 amounted to 11.2 ha, scattered all over the village area in fifteen fields, inefficient by any criterion of spatial organisation but highly successful as an example of territorial aggrandisement in an area of acute land scarcity. He had begun with a hectare, compared with Inuwa's half. Owing to his large household, he had almost the same land:labour ratio as Inuwa – exactly 1.0 – but this does not take account of his sons' private fields. On this large holding, he and his family kept 23 sheep and seventeen goats, and several fowls. He had two bulls and a plough.

Mikaila's fields are indistinguishable from Inuwa's in their general soil properties and cropping practice. This does not, of course, mean that the holding is homogeneous. Every area, however small, differs in its micro-ecology, in part because of the population of mature farm trees that covers the Tumbau area at a density of seven to twelve per hectare (Mortimore *et al.*, 1990). Some trees fix nitrogen, others do not; some cast heavy shade during the cropping period, whereas the common *Faidherbia albida* does the opposite, losing its foliage during the rains. Canopy characteristics and height also affect shading. Small ruminants graze leaves and pods under the trees during the dry season, and their preferences have an impact on nutrients in the topsoil.

Micro-topographical differences affect the retention of surface water after rain, and create small temporarily waterlogged sites where rice can be grown. Fertilisation, weed management and plough use in past years also affect soil characteristics. Poor management can initiate a cumulative cycle of poor returns, leading to further neglect which can only be broken by a large investment in manure and increased labour. The intricacies of managing a highly fragmented holding such as Mikaila's demand a high level of skill. As Netting (1993) has argued, family labour on such a holding is intrinsically superior to hired or cooperative labour, as supervision is embedded within the social interaction of the household.

Figure 9.2 shows how sharply different patterns of family labour management can ensue from different demographic endowments. Mikaila was able to maintain a smooth agricultural input throughout the farming year at little cost in effort (the highest rate of use achieved was little over 30 per cent of available labour), compared with Inuwa's inevitably fluctuating input and higher commitment. Furthermore, Mikaila's household's level of commitment to agriculture varied little from year to year, and the peaks were quite suppressed. There was ample capacity left over for his sons to commit to their own fields, and little need for the wives to work in his fields. This contrast underlines the vulnerability of the small household, constantly stretched, and exposed to the disruptive potential of sickness – or other unavoidable interruptions – even if affecting only one person.

Labour used in farm work was consistently higher on Mikaila's holding than that committed to livestock work (Figure 9.4). The relatively high commitment of labour time to domestic activities, both relative to farm work and when compared with Inuwa's household (Figure 9.3), indicates that women in Mikaila's house need not expect to undertake fieldwork, except (by convention) at sowing time and for the groundnut or cowpea harvests. Nor is it likely that his sons relied on women's labour in their own fields. Off-farm and business activities were suppressed, but resting occupied more time than in Inuwa's hard-pressed household, and sickness took its toll only during the later part of the season. After the harvest of 1993, Mikaila reported a surplus over his subsistence needs and planned to sell some grain, as he had done in the previous year (six bags).

The investment structure of this extended family is very complicated. The sons farm their own land, in addition to working on their father's, and the output belongs to them. But the ox-plough, which is used by all, was financed through selling groundnuts from the family land. Both Mikaila and his sons engage in buying, fattening and selling livestock, including plough bulls, in transactions by means of which they contribute to one anothers' needs. As the principal form of liquid asset, animals (which, though owned by almost every-one in the family, are managed together) are the currency of a virtual coop-erative bank which provides an effective cushion against personal need. Apart from Mikaila's small income as a *malam* (gifts and fees for religious duties), the family sees little need to diversify incomes further, though if there is a need, a son may be sent off to the city to trade, bringing the profits back to share.

Mikaila's situation suggests quite strongly that productive capacity can increase in response to a growing demographic endowment in family labour supply. Indeed, large households with little land appear to be rare in the Kano Close-Settled Zone. Young men, free from other commitments, can undertake *cin rani* and earn the capital necessary to increase the size of the holding. Such a household is in a strong position further to increase the productive potential of its labour force by embarking on plough ownership. It is also able to pool its capital resources for achieving common objectives. A question mark remains over the future, as the costs of acquiring land continue to rise. There is also resistance building up to alienating family land, even to neighbours. People prefer short-term arrangements such as loaning and pledging. Yet a large house-hold *must* maintain a favourable land:labour ratio if it is to be self-sufficient in food grains in most years.

Adamu Male of Dagaceri

Adamu Male of Dagaceri is an elderly man of more than 55 years who lives with his wife (aged 45) and her daughter by a previous marriage, who is 10 years old, in a small house in the main street. His own children have grown up and established independent houses, or married outside. With only 2.2 labour units, this is among the smallest households in the village (leaving aside widowed persons living alone). His household did not change during the four years between 1993 and 1996.

By the standards of Dagaceri, where landholdings tend to be much larger than in Tumbau, Adamu Male has rights to a generous amount of cultivable land – 5.6 hectares, which places him ahead of two other households who collaborated in our study. Adamu Male's land is in three fields, all inherited:

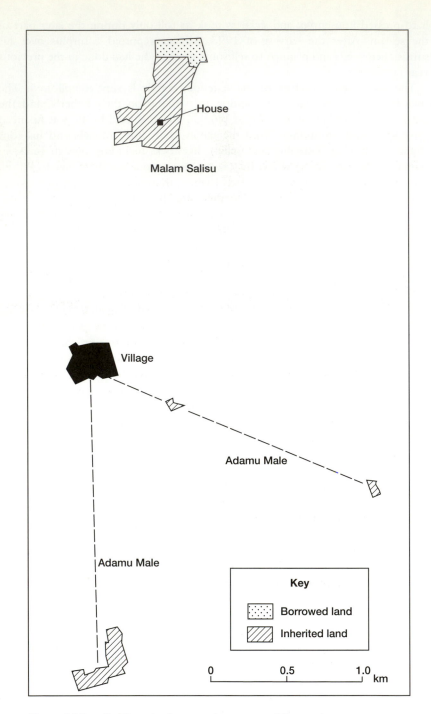

Figure 9.5 Landholdings in the extensive system of Dagaceri.

Plate 9.1 Dagaceri: *guna* melon in the fields after growing through the dry season.

two of them (of 0.4 and 0.5 ha) are in the old fields near the village, and a third, much larger (4.7 ha) is close to the eastern boundary where his father cleared and appropriated extensive land 30 or more years ago (Figure 9.5). With 2.5 hectares of land per labour unit, he would be hard pushed to cultivate all of it with family labour alone.

The soils of Dagaceri are coarse brown sands developed on former dunes, and are low in nutrients. Before the expansion of cultivation was curtailed by the reservation of grazing areas in 1972, fallowing was the preferred method of managing fertility. Today things are different. The old village fields are regularly manured, inorganic fertilisers are used where they can be afforded, and short grass fallows of between two and five years are employed on outlying fields. Farmers are, however, beginning to take the approach of their counterparts in Tumbau: a fallow is seen less as a fertility strategy and more as the result of a shortage of the inputs necessary to put it under cultivation.

Manure is incompletely privatised in Dagaceri, as most grazing takes place off the farm, in common rangeland or on other people's fallows; the communal middens on the village periphery are also open to all. Only pen manure can be composted privately, if farmers so wish. In any case, Adamu Male is in a weak position because his shortage of labour acts as a barrier to his adopting more intensive practices. He and his wife own six sheep (including a ram), but only one stays with them; the others are put out with a FulBe rearer nearby. This is also due to his labour shortage; so although he can reap some income benefits,

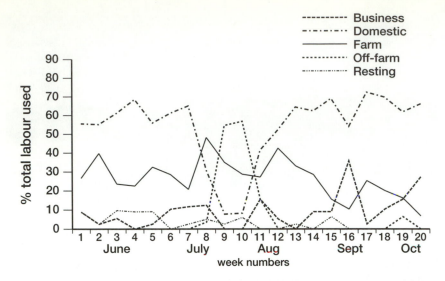

Figure 9.6 How labour was distributed between agricultural and non-agricultural activities in a small household (Adamu Male) in Dagaceri, 1995. (Labour use is shown as percentages of total labour used.)

there is no possibility of closing the cycle of nutrient management on his own land, in Tumbau fashion. A lot of his land is under fallow.

Adamu Male, with several other older men, makes mats from *kaba* shoots during the dry season and in the afternoons of the farming season. Although the returns from this work are low, it demands less effort than farming and can be scheduled on a regular basis, offering an 'incapacity option' to those whose farming days are nearly over. During the drought of 1972–1974 it brought in enough cash to enable many households to enter the inflated grain market and buy a little food.

How does Adamu Male manage his small labour force during the farming year? Figure 9.6 shows three peaks in farm labour in 1995 which were the three weeding cycles. In the interval between the third of these and the millet harvest, Adamu Male travelled out for two weeks on business which could not wait till after the harvest. During his absence his wife assumed full responsibility for farm work, and domestic work virtually stopped.

It would be easy to see Adamu Male as a victim of 'age poverty', in which a declining labour resource traps a small and ageing household into a cycle of declining output and under-capitalisation, at the same time closing off many opportunities for diversification. The circumstances in which he and his wife live are very modest. However, we have not yet told the full story. Adamu Male's son is one of the largest farmers in the village, having received all but four of his 22 hectares by gift from his father some years ago. This huge holding is managed using ox-ploughing and hired labour. Thus is his father's economic security assured in semi-retirement. We include this example in order to show how elusive poverty is when traced via standard economic indicators based on an assumption that households are independent units. Under variable conditions such as are exemplified in an extreme form in Dagaceri's harsh environment, answers to insecurity are, in fact, sought all the time through kinship, patronage or religious claims.

Malam Salisu of Dagaceri

In Dagaceri, the FulBe agropastoralists claim to have arrived before the Manga farmers, and their farms are very large when compared with those of most Manga households, and are situated in the interstices of the pattern of nucleated villages established by the Manga. The FulBe house (*ruga*) stands on its own land and usually within shouting distance of several others, forming a loose cluster under the leadership of the FulBe chief (*ArDo*) whose role includes the delicate negotiations often necessary with the Manga communities.

Malam Salisu inherited such a farm in the rangelands north of Dagaceri (Figure 9.5). His father died in 1996, but some years previously he had partly retired on a small fraction of the original holding, leaving most of it to his son on whom he depended for assistance from time to time. Malam Salisu had a household numbering ten persons in 1996. Still in the prime of life (aged 40), he now has three wives (aged 30, 29 and 23 – the last married in 1996). In the years 1993 and 1994, there were two daughters (18 and 16 years), a granddaughter aged 3, a niece of 14 and a nephew of 7. However, both daughters married and left in 1995; in 1996, his mother moved in with them after the death of her husband; and a second nephew of 20 years and another boy of 12 years moved in. This instability (falling from 4.8 labour units in 1993 and 1994 to only 3.1 in 1995, and recovering to 6.7 in 1996) is an important factor in economic management, as Malam Salisu struggles with a perpetual shortage of labour.

His farm of 14.6 hectares is all in one piece, and about six hectares are under cultivation in any one year. Some of this belongs (by his gift) to his wives, who control the use they make of their own plots. Small changes are made each year in order to rotate fallow, and in 1996 the house site was moved 100 m so that the manure left by tethered livestock around the old house could be taken up by the crops. Grain and grain-cowpea mixtures are planted; and in 1996 he used a little inorganic fertiliser (Harris, 1998). Some of the land suffers from infestation with the parasitic grain crop weed, *Striga helmenthuensis*.

Malam Salisu has an interest in livestock which far exceeds that of most Hausa or Manga farmers. In July 1994, the family's holdings numbered fifteen cattle, 29 sheep, 29 hens and his own flock of six guinea-fowl. Ownership was widely dispersed. His two wives (at that time) each owned some cows, sheep and hens, and his two daughters (who later married) were already set up with their own

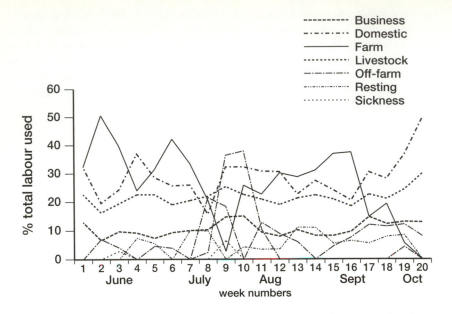

Figure 9.7 How labour was distributed between agricultural and non-agricultural activities in a large household in Dagaceri, 1995. (Labour use is shown as percentages of total labour used.)

cows or sheep. Even the 3-year-old granddaughter owned a couple of hens. We can see that the total mobilisation of the family is founded on sharing the livestock interest among everyone. The animals are, of course, managed together, most of the cattle and sheep being taken off each day, first to wells or pools for drinking, and then into the surrounding grazing area. Hen-houses stand in the corralling area outside the front door which, like all FulBe houses, faces west.

Labour management in the difficult year of 1995 (when family labour was scarce and rainfall was poor) is shown in Figure 9.7. The first important feature is the steady commitment to livestock work, which ran at 20–25 per cent of available labour throughout the farming year. By contrast, farm work fluctuated quite sharply, showing two weeding peaks followed by a slump, late in August, when Malam Salisu undertook a short absence from home ('off-farm'). Afterwards, the harvesting operations were prolonged and arduous, for the output from six hectares of cultivated land was substantial in relation to the number of people available to bring it in. During the season as a whole, the ratio between livestock and farm work was 1:1.4 on average. Malam Salisu worked long hours on the farm, and hired a plough team to help him on a large part of it. The curve for business work indicates the marketing of milk and butter by the women of the house, which entails visits to periodic markets as well as to Dagaceri village.

Of course, wealth is reflected in livestock holdings, which includes the condition and breeding capability of animals as well as their number. Malam Salisu did not rely only on rights of inheritance to set himself up. About 20 years ago, he worked for several years as a night guard for an expatriate household in faraway Lagos, accumulating resources for investment back home. Now he has neither the need nor the time to diversify. He believes that the fodder resources in the area can support more animals, and the constraints on further increasing his holdings are only financial. Like everyone else he buys and sells animals frequently, always looking for opportunities of long-term gain. His sustained optimism stands in sharp contrast to the impression of acute stress in the livestock sector which tends to be conveyed by expert diagnoses of overstocking and degradation of the natural rangelands.

Musa Manami of Futchimiram

Less complete data are available from our collaborating hamlet near Futchimiram, where Musa Manami (who is in his early thirties) lives with his young wife, and a son aged 7. In 1994, he married a second wife who was considerably older than her senior. His small labour force thus swelled to 2.7 for the year 1995 (no data were collected in the year 1994).

In the Badowoi parklands shifting cultivation is organised in collective blocks, fenced with thorn branches from the pollarded trees within, and rotated every few years as fertility falls, while the old fields close to the village are cultivated every year and manured with the droppings of domestic livestock. These old fields were set up, likewise, in a block when the fathers of the present community arrived about 40 years ago. They were subdivided into long strip fields as 'each man farmed westwards from his front door' with short eastward extensions behind the houses (Figure 9.8). The infields were themselves extended, piecemeal, westwards, as individuals added small square fields to their holdings. Musa Manami has divided his inherited entitlement to these in-fields between his two wives, each having approximately 0.6 hectares in two portions.

He confines his own work mainly to the outfields. When this study began in 1993, a block of land lying 1–2 km east of the village was farmed by half a dozen individuals, and Musa Manami had 3.7 hectares there. During 1994, falling yields on this land led to permission being granted for a block of long-fallowed woodland to be opened up, 1–2 km north-west of the village. In this block he now cultivates 4 hectares, having progressively abandoned the eastern field as its thorn hedge falls into disrepair. (As livestock graze without supervision for most of the year, the fences are a necessary condition of farming.) Pollarded trees stand at densities of three to eight per hectare, their branches removed in order to deny marauding flocks of *Quelea quelea* or other seed-eating birds a place to roost. Taking either the eastern or the north-western fields into account, Musa Manami's household has about five hectares of farmland, with a ratio of 1.7 hectares per labour unit. Mainly millet with some

179

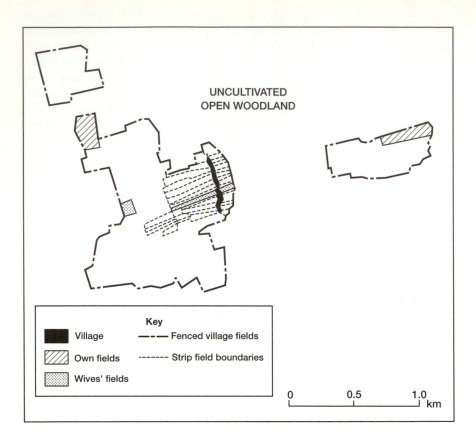

UNCULTIVATED
OPEN WOODLAND

Key

■ Village   —·—· Fenced village fields

▨ Own fields   ------ Strip field boundaries

▨ Wives' fields

0   0.5   1.0
km

Figure 9.8 A small household's landholding in the extensive system of Futchimiram
(Musa Manami).

cowpea is planted on this land, and groundnuts in smaller, closely planted
patches.

Like all Badowoi households, Musa Manami's owns livestock.[2] These include
cattle first and foremost, as well as sheep and goats. These mixed herds are
allowed to graze without supervision in the rangelands during the dry season
(when no crops are in the fields), even being left out at night. They also enter
the fields and graze the residue. But in the farming season they must be
supervised, and the village commits this work into the hands of two or three
children, who are recompensed. One of these is Musa Manami's son, although
he is only 7 years old. The animals are kept overnight in the corral at the east
(rear) of the house, whence their manure can be distributed to the nearby
infields – those of his wives. There is a horse (some households have two),
which is the accepted transport for longer journeys for both women and men.
Many houses also have donkeys, whose essential task, in addition to transport, is

180

Plate 9.2 Futchimiram: storing heads of millet in the *baga* after a good harvest, October 1994.

drawing water from the deep (40–45 m) hand-dug wells, on which both animals and humans depend exclusively from October or November (when the seasonal pools run dry) until July the following year. Transport animals may be watered twice a day, all the others only once. Because of their importance, donkeys are fed with grain and cowpea residues.

Musa Manami's little son works every day with the animals; his wives work on their own fields, and he (apart from trips to market) on his. The records of farm work in the two years 1993 and 1995 show a pronounced interval between the end of weeding and the beginning of harvest (Figure 9.9): an interval that was noticeably longer in 1995. In 1993 Musa got behind with weeding the groundnuts and had to obtain help, raising his labour force to 125 per cent of that available in the family, for a week at the beginning of September. In 1995 they managed to mobilise, on average, 63 per cent of their available labour for farm work (not including the son's livestock tending) in the eight weeks of the main weeding period. After the interval, the rate of use again rose to over 55 per cent during the prolonged harvesting season, which includes cutting, drying and storing the grain heads, felling the grain stalks and arranging them horizontally in the circular stook called *baga* in Kanuri, picking or lifting cowpeas and groundnuts, and drying and stacking their valuable residues.

The livestock attract most of the capital investment in Futchimiram; Musa Manami has no plough, nor does he invest in farm inputs. Land is not bought

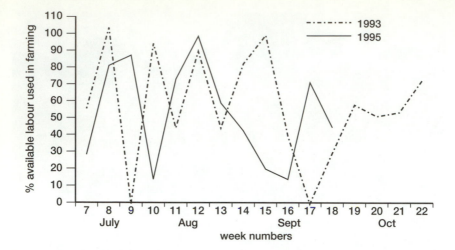

Figure 9.9 Percentage of available labour used in farming by a small family (Musa Manami) in Futchimiram, 1993 and 1995.

or sold. Even livestock are not traded as freely as elsewhere; they are sold mainly to meet contingencies, and their milk is consumed in the house. The groundnut crop and, if it grows, the vulnerable cowpea crop, provide cash for purchases. Although many visits are made to the small local markets this is a strongly subsistence-oriented system, with livestock playing the major role in wealth accumulation. Nevertheless, Musa Manami is in good standing with his neighbours as a hard-working farmer who achieves good results by the standards of this unremittingly harsh environment.

Kambar Madu of Kaska

Our final thumb-nail sketch is of a large household living near Kaska, well endowed with both labour and land. Kambar Madu is now an old man, but his two sons (aged 35 and 30) live at home with him. He has two wives (aged 50 and 40 years), his sons each have wives, and there are six grandchildren aged between 5 and 9, two boys and four girls. In addition, a woman relative lives with them. However, for arduous work, there were only two men (if he excluded himself) and a boy of 9.

Farm holdings are very large in the Kaska area. Farmers simply plant millet in the dune soils after the first rain, and appropriate the land by subsequent weeding. Boundaries are not built, except on the tiny *garka* irrigated plots in the *kwari* depressions. Thus Kambar Madu's main farm is an enormous field of 17.4 hectares to the east of the village, which climbs the slope from the lower *faya* soils to the upper *tudu*, where it is supplemented by another of 5.4 hectares. A third field of 3.9 hectares occupies *faya* soils on the north side

Plate 9.3 Kaska: *garka* plots being prepared for irrigation after the rains, September 1995.

(Figure 9.10). Some of this land is farmed by the women (his own and his sons' wives). No account is taken here of his small *garka* plot, on which most work is done in the dry season.[3] In total, about 26.7 hectares of rainfed land give a ratio of 3 hectares per labour unit.

A very extensive form of cultivation is necessary on these excessively free-draining, coarse, sandy soils, which have virtually no profile and very low nutrients. They support only a thin grassland of annuals (dominated by *Cenchrus biflorus*) under natural conditions, so to the casual visitor it is a surprise to see crops growing on them at all. Millet, cowpea and (after the millet) *guna* melon are planted very widely spaced. There are large germination failures owing to rainfall deficiency at the beginning of the season; rodents and grass-hoppers can devastate a crop before its seedlings have had time to grow more than a metre tall. On the upper *tudu* soils, on the dunes, a sort of shifting cultivation has resumed in which the fallow vegetation, being grassland, is vulnerable to wind erosion. More stable practice, however, is now found in the *faya* depressions, where the better soils are cultivated year after year, and sometimes manured. Weeding is vital to support the growth of the struggling crops.

The management of Kambar Madu's labour force during the farming year is very different from that of Mikaila's in Tumbau, even though they both have larger than average endowments of labour and land. Figure 9.11 shows, first, the high relative commitment to farm work that is necessary in the high risk

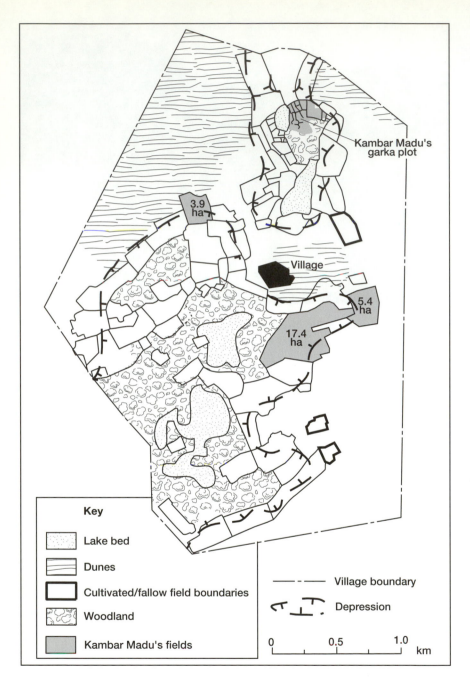

Figure 9.10 A large household's landholding in the diversified system of Kaska (Kambar Madu).

Source: After Ibrahim (1996)

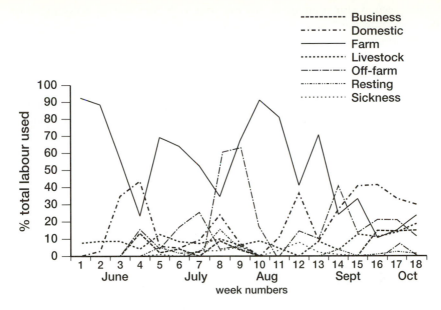

Figure 9.11 How labour was distributed between agricultural and non-agricultural activities in a large household (Kambar Madu's) in Kaska, 1995. (Labour use is shown as percentages of total labour used.)

environment of Kaska with its compressed growing period. From mid-June until the beginning of September 1995, 64 per cent of the labour used, on average, was committed to farm work in each seven-day period (reaching 90 per cent in two weeks), compared with 30–40 per cent for Mikaila of Tumbau. Yet this work fluctuated widely, displaying three weeding peaks before September, whereas in Tumbau we observed a rather smooth curve. From September, the labour committed to harvest work by Kambar Madu's household dropped off quite sharply as output was disappointing after poor rainfall.

A second feature of the graph is a reciprocal relationship between farm and domestic work, whose respective peaks and troughs complement each other. This is because women's labour was diverted from domestic to farm work at each of the three peaks, and only recovered after harvesting work had assumed its decline. Livestock work was quite consistent throughout and rather low, because we weighted children's labour adversely. Business and off-farm work were suppressed until the important farming tasks had been completed. All in all, the picture suggests a furious effort, calling for everyone's commitment, to capture as much economic value as possible from an all too brief episode, during which an erratic rainfall mobilises precious nutrients in the soil.

Interviewed after the harvest of 1993, Kambar Madu reported that the household expected to have a small grain surplus over the year's requirements

185

for the second year running. In the previous year, four bags had been sold by the women in the house, and they would sell again in the coming dry season.

Opportunities for income diversification are better in Kaska than in either Dagaceri or Futchimiram. Most men can produce market crops in their *garka* plots during the dry season without having to leave the village, or to divert labour from subsistence grain production – though it must be said that transport to markets is slow and expensive. Or they can enter the trade in *kanwa* (natron), which is collected by trucks from Gashua. The *kwari* depressions have groves of date palms whose product is readily marketable. And there is the animal trade. FulBe stockbreeders in the area complain that owing to frequent droughts, the terms of trade between livestock and farm products often move in favour of the farmers, who are said to be increasing their ownership of cattle in particular. As the farmers usually commit their cattle to FulBe stock keepers for grazing and management, such a transfer, if it is occurring, tends to be invisible. But Kambar Madu participates in cattle ownership, and the livestock market, as elsewhere, offers him an opportunity to better his holdings.

This area is famous for the visible evidence of 'desertification' in the form of recently reactivated sand-dunes, which cover more than 20 per cent of the surface, disappearing woodland, and declining water-tables (Mortimore, 1989; Carter, 1994; Reenberg, 1994; Ibrahim, 1996; Alhassan, 1996; Turner, 1997). It is quite difficult, however, to link such processes directly with the livelihoods of specific households. Kambar Madu's shows that it is still possible to prosper, given enough rain to grow the all-important millet.

What is poverty?

Something of the elusiveness of poverty is suggested by these sketches of individual households in our four villages. It is as difficult to pin down the 'poor household' as it is easy to generalise about poverty in the Sahel as a whole. In the literature about poverty in the West African savanna and Sahel, it has so far proved impossible to demonstrate a class basis for rural impoverishment, as in India or Latin America, for example. We suggest that this failure is due to the fact that even taking a strictly economic view of poverty (that is, ignoring its social dimension), there are too many variables involved. In Figure 9.12 we identify six that we consider to be the most important which, besides contributing directly to a household's relative wealth or poverty, interact with one another in the following sequence:

1   the labour resource with which a household is endowed, and also its future labour resource (anticipated in having children, especially males, growing up in the house);
2   the stock of farm capital, such as ploughs, manure or improvements which, with labour, determine productive capacity;

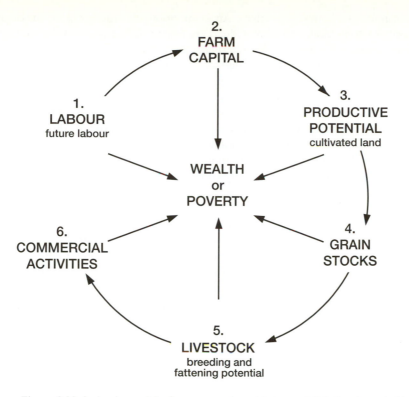

Figure 9.12 A simple model of poverty and wealth in rural Sahelian households.

3 the productive capacity of the cultivated land to which the members of a household have rights of access (not necessarily by ownership);
4 the grain stocks held by the household after the harvest is over, which determine the security of the household's subsistence in the coming year;
5 the livestock owned (or perhaps managed) by members of the household, which are used as savings, and their fattening or breeding potential, which is interest on those savings;
6 a commercial portfolio, comprising business, manufacturing, service or distant options enjoyed by members.

If this is a simple model of *local* causes of wealth or poverty, it must be understood to sit embedded in a larger economic system which dictates, from the outside, many of the terms – for example, access to land is affected by tenure legislation, access to markets is affected by prices, commercial activity is affected by controls, and development project interventions alter the status quo. If it is also a restricted *economic* model, it must be understood to operate within a specific *social* context which affects the ways in which people can respond to opportunities to better themselves. An assumption that there can be an open system, at village level, where individuals are free to pursue their own advantage without social constraint is almost certainly unrealistic, even today.

Such a model should not be understood, either, to imply any kind of determinism in outcomes. The cycle of interactions which it proposes is, at every stage, directly affected by management decisions on the part of individuals which include the usual proportion of mistakes, incompetence and bad luck as well as strokes of genius. In such a multi-variate process of economic differentiation, the effects of psychological, pathological, and temperamental factors as well as the influences stemming from kinship, patronage or other social networks and pressures are very complex, and introduce a random unpredictability into the data when it is aggregated. One reason for caution in discussing poverty in such a context is the existence (though generally underestimated in unidirectional theories) of countervailing mechanisms whereby individuals and households seek to climb back from disadvantaged positions, and sometimes succeed in doing so. 'Chayanovian' cycles – whereby the demographic growth of the family is linked with an expansion of landholding rights, later to be succeeded by dissolution and a fresh start by the next generation – have been claimed to work in this way in northern Nigeria (Mortimore and Wilson, 1965; Hill, 1972, 1977).

If we must search for a single indicator of wealth or poverty, *livestock ownership* has four arguments in its favour. First, in all of our Sahelian villages, livestock ownership is deeply rooted in an indigenous system of values which places priority on animals as a form of savings. As they multiply, and can be increased in value through fattening, they are productive investments, and not a sink of unproductive wealth as they were so often represented by uncomprehending officials in the past. Second, the observed behaviour of rural households shows that animals are mobilised to meet contingencies; and what better indicator of wealth – in a high risk environment – than always having the means to deal with contingencies? Third, they have important feedback effects on farm productivity, effects which are gaining new strategic significance in the light of the necessity to intensify with low external inputs. Finally, they are used by many – if not most – owners as a means of increasing personal wealth through market dealing, by buying and selling when prices move. Among the variables identified in Figure 9.12, therefore, livestock assume a key role, as they provide the currency for savings, insurance, investment and speculation. Far from being the harbingers of environmental destruction, as portrayed in much of the literature about desertification (Chapter 4), livestock are better seen as a measure of a household's capacity to sustain livelihoods and accumulate wealth in the face of environmental and economic risk.

10

MANAGING THE MANAGERS

Karambanin akwiya, gai da kura
(*It was meddlesome of the goat to greet the hyena*)
(Hausa proverb, Abrahams and Kano, 1949)

Beyond the 'typical' household

This has been a study of smallholders, a class of householders whose distribution and importance extends beyond north-east Nigeria in particular, and the Sahel in general. Netting (1993) has argued the case for recognising such a productive mode, whose distinguishing characteristics include autonomous resource management, private property rights, mixed subsistence and market production, and the use of family labour. Smallholders, according to this conception, are found not only in post-colonial West Africa, but in the records of sixteenth-century Switzerland, or in China from the Han Dynasty to the post-Maoist 1990s. Among a diversity of farming systems, the theme which is common to all is an historic push to intensify agriculture, in response to a need to feed and employ more people. West Africa is a classic locale for studies of smallholder systems, among which Netting's own study of the Kofyar (1968) holds an honoured place.

A desire to quantify, which has long been accepted without question as a prime objective of farming system studies, can have a perverse effect on our understanding of smallholders. Farm input–output analyses quantify the factors of production (land, capital, labour) and estimate profitability as a basis for development promotions of new technologies, infrastructures or subsidised inputs. The assumption that underlies such analyses is that the smallholding may be understood as an agricultural micro-firm, which differs from its opposite – the large (commercial) enterprise – mainly in scale. Given such an assumption, it is rational to advocate labour-saving and yield-enhancing innovative technologies, in order to increase household or individual incomes from agriculture.

Why have so many agricultural developmental initiatives failed or, at best, enjoyed only a qualified success in smallholder economies? Why have rural development interventions often benefited only a few at the expense of many, accentuating existing disparities of income and welfare, instead of removing them? Why have new technologies so often enjoyed only a slow and uncertain uptake, even when thoroughly researched at the farm level? Why do smallholders often persist in land use practices, crop preferences or livestock

189

ownership patterns against the advice of extension services? Answers to these and similar questions can be sought in three directions: first, in the diversity, second, in the variability, and third, in the multidimensionality of smallholder systems.

Diversity

The quantification of smallholder farming systems is usually directed towards a search for the 'typical' or 'representative' unit, which can be used as a basis for diagnostic and prescriptive development interventions. While it is commonplace to point out that the 'average' scarcely ever exists in reality, nevertheless, the full implications of using statistical averages in a population as diverse as our four communities (for example) are rarely if ever recognised. Their diversity on any chosen parameter (population, family labour force, land: labour ratio, capitalisation, enterprise mix, etc.), and their fluidity from year to year, have been emphasised in the preceding chapters. It may be questioned, too, whether the concept of the 'normal distribution' also disguises variability, in the sense that it is confined to one variable at a time (where many variables interlock), and leads to extremes being treated as deviant. Rather, the scattergraph more appropriately describes the diversity with which we are faced. The problem is not merely graphical. If a 'normal' smallholder cannot be found, how can appropriate policies be prescribed?

Variability

Unlike large farms, smallholdings can vary greatly within a small space or over a short period of years. Many smallholdings (as our case studies in Chapter 9 show) consist of a diverse and scattered portfolio of fields. Even within a single holding there can be sharp differences in input management and in productivity, not only by accident, but also by design. Environmental diversity is matched by management diversity. Soils are managed on a patch-by-patch basis, their different responses to inputs being well known to the farmer. Animals are also tended as dissimilar individuals rather than as a homogeneous herd, in complex herd structures designed to achieve multiple economic (and social) objectives. Commercial farms vary within rather narrow limits which represent acceptable profit margins of particular crops, animals and regions; to transgress these limits is to compromise viability. Smallholdings, on the other hand (as market production may not be a dominant objective), can afford to vary more widely without necessarily becoming uneconomic. In a village of smallholders, there is a place for bad as well as for good farmers (or livestock producers), for both competence and entrepreneurial ability vary. From year to year, too, the same types of variability occur.

Multidimensionality

This is not all. Not only are smallholdings incorrectly conceived as 'micro-largeholdings', but they are qualitatively different in their multidimensionality. Far from being summable in an equation of the factors of production in farming, they are driven also by a need to diversify economically. We have represented this, in Chapter 7, in a stepwise model. In this model, economic and social behaviour meshes, and the household is embedded in the community matrix. It is a mistake, as has often been shown, to regard smallholders as driven at all times by agricultural profit margins, even though there are many examples documented wherein they have responded with speed and flexibility to market opportunities. In such responses, as in all else, there is at any one time a continuum from apparent failure (impoverishment) to apparent success (wealth). However, fortunes can and do change, sometimes quite rapidly (for example, with the death of a key household worker or a drought), and sometimes slowly (for example, with passing generations).

If these inferences are correct, the policy implications are somewhat startling. Questions must be raised concerning the suitability of strategies targeted at the 'normal' or 'average' smallholder, under 'normal' conditions, and equally fundamental queries should be raised about the validity of concentrating development efforts on the farming sector (and usually on one or two selected crop enterprises at that) in systems which thrive on their sectoral diversity and multiple objectives. The most impressive stories of development are those where a need for multiple choices, to suit a range of smallholder families, has been met, implicitly or explicitly, in the type of interventions and opportunities affecting rural households (Tiffen *et al.*, 1994).

The argument of this book

Although this study suggests a resonance with the experiences of smallholders elsewhere and at other times, it is not about *all* smallholders – on whom, as the foregoing makes clear, it is impossible to generalise – but a particular category of smallholders, those inhabiting African drylands. The distinguishing features of drylands are aridity, environmental variability and low agroecological potential. Within this category of African dryland smallholders, we have explored the situations in which smallholders of the Sahel find themselves. A distinguishing challenge facing Sahelian smallholders is their subjection to a decline in rainfall (which is expressed in increasingly frequent droughts) over the past three decades. This environmental change is obviously central to the experience of farmers and livestock keepers in the Sahel. The diversity, variability and multidimensionality which characterise smallholders as a whole take on special significance in this Sahelian context.

In Chapter 2 we laid a foundation for a systematic analysis of the problem, in the form of a model linking diversity, flexibility and adaptability in a Sahelian smallholder system, which is subject to high levels of uncertainty in both the natural and the political-economic environment. We sought to uncover diversity by adopting a multi-sectoral approach to the complex household system, and a multi-site sampling design to uncover a range of conditions in ecology (rainfall), demography (population density) and economy (market access). We sought to uncover flexibility by monitoring labour management (in particular) on a continuous basis throughout the farming year, and household by household. Finally, we sought to understand adaptability in the longer term by prolonging our observations over four successive years, in which rainfall conditions and the circumstances of individual households varied.

Beginning with the flexible management of farm labour in response to rainfall, we have shown in Chapter 4 that inter-annual variability, and the compression of the growing period which usually accompanies a reduction in rainfall, impart an opportunistic flavour to farming goals and practices. It is very different from smallholder farming in equable climates, and from large-scale farming in which commercial goals, technologies and insurance systems can go far to cancel out the effects of climatic variability.

A diversity of responses, both between places and between years, calls into question a conception of 'Sahelian agriculture' that is circumscribed by certain averages, normal distributions, and normative prescriptions based thereon. Certain 'inefficiencies' are necessary conditions of success in managing such a difficult environment. First is a need for some 'surplus' farm labour in dry years in order to maximise output, and control bioproductivity, in wet years. Second is some unavoidable idleness during the long wait for the first rains, when all hands are needed within 24 hours. Third is a need for 'reserve labour' at the household level in order to cope with the unpredictable nature of the farming year. Finally, fourth is the well-known mismatch between employment in the dry season and the demand for labour in the growing period.

The more compressed the growing period, the higher the percentage of available labour used in farm work, accentuating the risk in marginal environments. However, with increasing rainfall, an increased scope for intensification brings about an increased amount of labour diverted into livestock management. Labour deficits translate into poor weeding, inadequate thinning, poor growth, unnecessary harvest losses, reduced yield and lower output per worker. Farmers are aware of such losses, but opportunistic farming offers potential benefits from planting very large areas – at least when rainfall is good. Thus 'waste' of effort, of seed and perhaps sometimes of fertility are an integral part of the farming system. 'Efficiency' would leave no room for flexible manoeuvre.

Harvesting labour patterns show that major activities are separated from one another by transitions when the 'terms of labour' for competing tasks switch values. Negotiation of these transitions calls for skilled management of the household labour force (and extra-household labour, if available), and their anticipation, by choice of appropriate planting dates.

Clearly, the idea of an 'average smallholder' throwing down a standard package of 'traditional' farm technology against the Sahelian elements is quite inappropriate. Development projects or policies have focused in the past on economic or technical constraints facing African smallholder agriculture. 'Magic bullet' solutions such as irrigation, or 'green revolution' packages, or mechanisation, have been favourites of the development community and national governments in Africa, but have a poor record (e.g. Williams, 1981; Richards, 1985; Adams, 1992). More attention could, with justification, be given to understanding the intricacies of negotiating a way through the uncertainties of Sahelian rainy seasons, given the capital and technological constraints.

In Chapter 5, we reviewed the ways in which labour has been employed to transform natural landscapes into domesticated ones. Using accumulated knowledge and skills, these communities have learned how to manage *natural* plant biodiversity, which they value, to their own advantage. The place of *crop* biodiversity is critically important in Sahelian farming systems, for which no 'miracle' seeds exist. In order to understand the continuing process of domestication better, we introduced molecular genetic analysis to pearl millet, the staple food crop of the region and an ancient, indigenous cultivar. This approach quantified the impact of farmers' management of seed stocks on genetic resources, suggesting that even while a measure of diversity protects the farmers from environmental risk, their unrelenting selective pressure maximises the value of their output. Crops, however, are not the farmers' only resource; we thus reviewed their management of fodder resources for livestock, and their transformation of natural woodland into farmed parkland, which contains both a valuable timber resource and source of harvested tree products.

The outcome of these activities is a transformation of the landscape (measured in terms of land use and vegetation classes) which, traced over time and compared between villages, shows a diversity of resolutions to the challenges facing the managers of natural resources. These resolutions are in fact investments, in biological and technical knowledge, skills and labour time, which have been made for the most part in small increments, and by large numbers of individuals and households. Nothing could be further from the scenario of reckless resource degradation which has been put about by some academics and development agencies. Supporting these investments should be the aim of development interventions, rather than disrupting centuries of incremental smallholder investment with massive public sector investments, whether (like dams and irrigation schemes) aimed at dramatic shifts in production, or (like afforestation programmes) aimed at equally dramatic changes in landscape ecology.

It is perhaps paradoxical to claim that the Sahel, for long thought to be victim to degradation induced by humans, is capable of sustainable management with enhanced productivity. However, in Chapter 6 we challenged the notion that the Sahel has overshot the limits of its productive potential, and investigated ways in which smallholders have sought to intensify farming.

Each of the four farming systems has a regime of fertility management which reflects the special conditions of soil, water, labour and need that characterise its neighbourhood. No farmers use inorganic fertilisers on a regular and sustained basis. We examine intensification, therefore, not as a technical option for the future, based on the use of inorganic fertilisers, but as an indigenous process. This provides a more realistic basis for evaluating its potential in future.

We began by describing the soils and their management in each of the four study areas, showing that in each, a specific set of solutions has been sought to the fundamental problem facing all farmers everywhere – that of maintaining the productivity of their soils. Then we asked whether an indicator of land use intensity (the cultivated fraction) is positively related to population density. While confirming such a relationship in space, we found no clear trend through time, and also that the value of such a simple model is restricted by the chosen indicator and by situational diversity.

We then approached the measurement of labour use per hectare directly, finding a predictable relationship with population density when *agricultural* labour inputs are averaged over all productive land, but not when farm labour inputs on *arable* land are considered. While the high intensity of the Tumbau system is clear beyond doubt, in the lower intensity systems the data must be understood in their systemic context.

To sidestep some methodological difficulties, it is possible to measure, in place of total labour use, peak labour use per hectare in key farming operations. The use of this refined indicator can facilitate comparisons both between systems and between years. It also permits the interactions of labour intensification with rainfall variability and other ecological controls to be clarified.

These analyses reaffirm the diversity and complexity of a process that has been the subject of much abstract generalisation. Given these complexities, and methodological difficulties, it is clear that the term *intensification*, when reduced to the farm level, covers more than one meaning. Given the diversity and variability highlighted above, it is not surprising that the decisions made by households do not always fit into a mould defined for them by overarching theory. All, however, are encompassed by a drive to sustain productivity from finite land resources, which gives the lie to a facile scenario of environmental destruction.

Sahelians are not only farmers (crop producers). In Chapter 7, we broadened the basis of our investigation to take account of the diversification of the household production, income generation and social system. At this point the weaknesses of regarding smallholders as micro-firms in farming become more obvious. At the same time, their exploitation of diversity, their flexible use of labour and year-on-year adaptability are continuing themes.

We have pursued farming households, when they are not farming, through three steps of diversification. These steps are: first, livestock husbandry; second, 'trading, making, and serving'; and third, the 'off-farm' activities exemplified by labour sharing, schooling, marketing and travel. The categorisation corresponds to significant shifts in the relations between family labour and the other factors of production, land, capital and skills.

In crop husbandry, all the factors of production are combined for primary production, using the cultivated land to which members of the household enjoy rights of access. In the first step of diversification, output from the secondary producers (animals) is added to that of crops, access to additional natural resources (such as common rangeland) becomes more important, and there are synergies in the form of manure and animal traction which benefit the crop-producing enterprise. In the second step, 'business', cultivated land assumes a marginal importance, though common property resources (such as wood, herbal medicines or water) may increase; however, the capital resources and skills of the household mainly determine the options available. Profits from these activities can be (and often are) returned to the crop-producing enterprise as investments. Finally, in the third step, what we have called 'other off-farm activities' may depart significantly from the household's own endowment of natural and capital resources and, depending essentially on skills (knowledge, information and networks), proceed to exploit opportunities elsewhere. Profits from these – where economic transactions are involved – can also be returned to the other levels.

In Chapter 8, our objective was to move from the analysis of household labour use in the aggregate to focus on the contributions made by women and children to securing household livelihoods, in particular in farming during the season.

The inclusion of women and children in the analysis increases still further the complexity and diversity of the subject. An analysis of how women's labour, and that of children, is mobilised in the context of the household's full range of interests indicates a diversity of solutions, which are affected by the rainfall (its amount, variability and peak distribution), by scarcity of cultivated land, by women's access to land, by the practice of seclusion and by customary division of labour. Women's labour, and to some extent also child labour, is essential for maintaining the flexibility of the household's response to variable conditions. Especially in places where married women do not take part in many field operations, they are a labour reserve which can expand the labour force in time of need.

These solutions should be understood as the outcome of a process of negotiation that goes on continuously. In this process, the internal structure of the co-residential household is important. It is not simply an authoritarian hierarchy, subject to the control of the (male) head. Such a model was always inaccurate (though many decisions are made in this way). Neither is it accurate to disaggregate the household into its individual members and to pit women against men in a competitive struggle. The truth lies somewhere between. What is clear is the functional effectiveness of the negotiations that take place. These smallholder-based farming and livelihood systems are demonstrably viable and resilient. They are actively engaged in a daily and seasonal search for livelihood security, wholly based on what has come to be labelled 'sustainable' systems of production.

Sickness, and resting, were also measured. The data show that sickness can make a significant impact on labour availability at critical times of the season, owing to its tendency to occur irregularly, especially during the middle of the rainy season.

Finally, in Chapter 9, we presented selected case studies of how households in different areas and circumstances sought to resolve the challenges facing them during the period of our study. The force of constraints, and the horizons of opportunities, interlock with the labour, land and other resource endowments enjoyed by households, creating a continuous differentiation in society. Recurring accidents of birth, death, inheritance, rainfall and a host of other contingencies ensure that the social kaleidoscope is frequently reshaken. How a household uses its resources of diversity, flexibility and adaptability – its management skills – helps to determine the outcome. Rigid models of inter-household differentiation, predicated on exogenous driving forces, cannot account adequately for the variable and fluid patterns of inequality observed.

The case studies allow us to identify six key constituents of household wealth (or poverty): (1) the labour resource itself (and its future potential), (2) farm capital, (3) the bioproductive potential of cultivated land, (4) the household's grain stocks, (5) livestock holdings and their productive potential, and (6) the 'commercial portfolio' of household members. The multidimensionality of poverty in Sahelian households is, according to this view, a major reason why it is difficult to ameliorate. If a single indicator of wealth is thought essential, we suggest that livestock ownership is central. Indeed, the findings of this study suggest that the role of livestock should be re-evaluated. However, the local processes of economic differentiation are embedded in a social matrix at the community level, and both are subject to strong exogenous forces emanating from political, social and market structures in the national and global systems.

We have, in this way, attempted to follow through the implications of diversity, variability and multidimensionality. The Sahel is a diverse environment, and the matching diversity of solutions sought by Sahelian households to its challenge has not been adequately recognised in the past. The Sahel is also a disequilibrial environment, and it is necessary to recognise that Sahelians have evolved adaptive strategies to deal with its consequences, and will not be helped by developmental promotions that push them towards specialisation or new forms of dependency. Furthermore, adaptation must continue in the future, as the idea of a 'miracle crop' for the Sahel is a chimera. The livelihood systems are characterised by a multidimensionality which has likewise been neglected, yet remains a potential resource. The complexity of adaptive behaviour requires a holistic approach to development, in sharp divergence from much earlier practice.

Adaptive policy

It is widely agreed that the Sahel has been a graveyard of well-intentioned development projects, although the latest signs are that a renewed focus on small-scale, bottom-up, participatory projects based on proven introduced or indigenous technologies is turning the tide (e.g., Reij *et al.*, 1996; Rochette, 1996). The great majority of interventions have attempted to address agendas defined from the outside by governments, professionals and international agencies. Prominent among these have been soil and water conservation objectives.

If the argument of this book is accepted, it follows that to base interventions and development policies on the assumption that Sahelian livelihood systems, and systems of natural resource management, have been tried and found wanting, is mistaken. Where longitudinal studies of dryland management have been carried out, the superficiality of the 'diagnostic-prescriptive' mode of development traditionally practised by professional arms of governments and agencies has become clear (Tiffen *et al.*, 1994). To take an ahistorical view is to ignore a need to establish trends (whether positive or negative), evaluate indigenous management over time, and understand temporal variability. It is not surprising that such an approach too often ended with technical recommendations based on expert advice, which were never assessed in economic or social terms.

Meanwhile, Sahelians have not been idle. Increasing rural population densities and market production of food crops, improving national food sufficiency, and intensifying interaction between town and country have not been confined to Nigeria (Snrech *et al.*, 1994). Intensive farming systems in areas of high rural population density have not failed, as has often been predicted. Rising land prices and 'scrambles' for farms, even in the driest areas, indicate that rural people (and many urban ones also) have not written off their productive potential. Observers in the francophone Sahel are carefully watching northern Nigeria, where urbanisation and monetisation have proceeded furthest, with interest – notwithstanding the failures of economic planning and mismanagement that have characterised its successive military governments. Can our four villages provide any pointers to the way other parts of the Sahel will go in future?

These four villages have shown us the scale and detail of the diversity which exists in the Sahel, and which needs to be taken into account by development agencies. They have shown the flexibility that characterises the management of resources, and especially of labour, and the adaptability from year to year. The twin policy goals of alleviating poverty and mitigating environmental degradation will best be achieved through enabling such communities to mobilise more technical, management and capital resources, to realise autonomous solutions to their specific problems. An adaptive policy framework which aims to support rather than direct – enable rather than control – positive change in a Sahelian environment should be the aim, and it needs to be based on empirical analysis, and to be less reliant on ill-researched degradation scenarios. In addition to meeting technical objectives, therefore, we can suggest the following nine principles for inclusion in a policy framework:

1   *Countering variability.* The complex interaction that has been shown between rainfall and labour management is a major determinant of agricultural output, and development interventions (for example those promoted by rural development programmes) should be designed to address this interaction, rather than merely productivity constraints, if they are to be inserted sensitively into the system. An important dimension in such a strategy is protecting small, household investments against the impact of variability in rainfall or against other hazards. Investments that have to be liquidated when food scarcity strikes the household are not available when the rainfall comes back; and growth cannot be resumed until rehabilitation has been achieved. The 'ratchet effect' of droughts on household poverty, unless it is confronted, will continue to threaten the continuity of any development programme. More work is needed on formally integrating risk and variability into development planning for poor households.

2   *Promoting diversity in adaptive technologies* through smallholders' own selection, experimentation, adaptation and combination of technical, management or genetic solutions. Research and extension (based on farmers' and stockraisers' participation) should be inserted into a broad framework, including indigenous technical knowledge inherited from an earlier generation as well as new solutions. Multiple sources exist to be tapped. Informal sources and travel can play as important a role as public investment in the race to adapt Sahelian agriculture to changing conditions.

3   *Facilitating the flexible use of labour.* Labour management is the key to successful utilisation of rainfall resources. There is a major differentiation to be made between short season systems requiring sharply peaked labour regimes and longer season systems where the sequencing of multiple operations is more important. The higher the percentage of available family labour that is used, the less flexibility is available to households which cannot gain access to non-family labour, and to the system as a whole. In view of the importance of these observations, development effort in agriculture needs to place a new emphasis on finding ways of assisting the efficient use of labour.

4   *Enabling agricultural intensification* through closer integration of crops wih livestock. Here we emphasise *enabling* indigenous intensification rather than *promoting* particular technical packages (Breman, 1990; McIntire *et al.*, 1992; Bosma *et al.*, 1996). The diversity of the intensification process – too easily caricatured as pumping inorganic fertilisers into the soil – requires recognition, in particular in respect of: (a) the specific characteristics of systems (rainfall, soils, cultivar inventories, etc.); (b) the ways in which fallowing is operated; (c) the changing ratios of intensively to less intensively used land; (d) access to arable land. Development agencies cannot

influence the intensity of land use, as it depends on labour management decisions taken by individual households. These in turn depend on factor ratios at the farm level at a given time. This explains why attempts to force the pace of crop–livestock integration have sometimes been disappointing. They should not interfere with land or resource tenure, since experience shows that such interventions obstruct flexibility, and redistribute benefits in inequitable or unforeseen ways. What they can do is to help extend the range of technologies available to farmers who are working with a land constraint. The responses to land scarcity identified in this study show the importance of diversity of choice under an intensification imperative. As crop–livestock integration is associated with intensification, new technologies or modes of management should integrate systems rather than addressing only one component. Research should target improvements in the efficiency of nutrient cycling, rather than identify theoretical capacity ceilings.

5   *Multisectoral scope.* The data obtained in this study indicate that agricultural production (crops and livestock) should take its place alongside households' non-farm labour commitments in agendas for rural and community development programmes, and suggest that a broader developmental framework, if adopted by governments, will benefit a wider social spectrum than addressing only questions of agricultural productivity. Development interventions, and policies designed (for example) to strengthen the informal sector, need to be based on analyses of the diversity in the rural non-farm sector, especially in the Sahel where diversification represents a key resource for dealing with rainfall variability. Because diversification brings Sahelian households into the regional economy (through migration and trade), open borders and unrestricted travel, especially on a north–south axis, will contribute to economic security.

6   *Promoting open market conditions.* The study has shown that even 'remote' communities are integrated into national and regional markets for food and other products. Urbanisation is driving these markets. Policy priorities are therefore improving farmgate prices, and transport and marketing infrastructures. The regulation of inter-district or international movements of food is counter-productive. At the same time, it must be recognised that more people, activities, commodities and factors of production are entering local markets and exchange systems, both in response to risk and in the hope of accumulating wealth. The response of farmers and livestock producers to markets is now better known as a result of many agricultural economic studies, and strong, long-term trends in market hinterlands (Snrech *et al.*, 1994) support a view of Sahelian smallholders in general as responsive to new opportunities. In place of the export commodity markets of the past, local and rural–urban systems of circulation have an increasing role to play as engines of growth.

7   *Alleviating poverty among vulnerable groups: poor households.* Small households are more stressed than larger ones, as they have less flexible scope for

the use of their labour. Poverty alleviation programmes should take account of both the weakness of small households' resource endowments and of their operational difficulties. Given the importance of labour management, demonstrated in this study, households with a smaller endowment are relatively disadvantaged, with negative consequences for their output and incomes. Defining poor households as small households is an oversimplification. But, as demographics are linked with social differentiation in the community, more policy research is warranted on equity and practical issues.

8   *Alleviating poverty among vulnerable groups: women.* Interactions between farm and domestic work concern, in particular, women's participation in farm work, and cultural practice which is specific to communities. The large amounts of labour time put into domestic work, and the competition that is implicit between farm and domestic work, underline the need to identify development objectives in this area. Development interventions intended to benefit women farmers require a basis in culturally and ecologically specific data and quantitative analysis such as that provided for our villages. Where they have access to more land, they have more autonomy in decisions concerning its use. Caution is warranted.

9   *Reducing the impact of sickness.* Time lost to sickness is quantified in this study, although our data do not allow analysis of specific medical conditions. Given the critical role of labour in the production systems, the high percentage use of available labour at certain times, and the arduous nature of farm work in high temperature and humidity conditions (which bears on all, irrespective of age or sex), *non-work* (both rest and sickness) has both economic and nutritional significance. It is an area where more information is needed as a basis for intervention.

In search of sustainability

The adaptive capacities of rural communities and their systems of production and income in the north of Nigeria have been forged over centuries of occupance in one of the world's harshest inhabited natural environments. Lacking a long-term perspective, preoccupied with equilibrial notions of human ecology (and the 1970s' environmentalist urge to 'save the planet'), and impressed by the destructive impact of drought, the north with its media drew some rather over-hasty conclusions in the aftermath of the Sahel drought (Adams, 1990; Mortimore, 1998). The environmental challenge to human adaptability was portrayed in great detail, but the resources of its communities to meet this challenge were ignored. A quarter of a century later, it is time to review those conclusions in the light of what has been learned.

To take an unfashionably optimistic view of the current management of the Sahelian environment and its future is not to deny the evidence that exists for degradation, nor to underrate the poverty that confines most of its inhabitants to levels of living that are unacceptable on grounds of natural justice. But to ignore the propensities of Sahelian systems to cope with variability both in the short and in the longer term is to deny them a future, except perhaps as passive recipients of improbably massive aid. These Nigerian communities, with their systems of resource management, point towards possibilities of transition to more productive, conservationary systems in future, given the right conditions. Pessimists cannot claim that the scope for such transitions has been exhausted; too little is known even to guess at the future trajectories of labour intensification, soil fertility management, technical change, crop–livestock integration and economic diversification.

We have gathered together these responses in terms of a simple model linking diversity, flexibility and adaptability in the household. Because we have focused on variability and change in the natural environment, it should not be thought that we diminish the importance of uncertainty stemming from impulses in the economic or political environment. In Nigeria, during the four decades since independence, there have been plenty of these. Nor do we underestimate the force of global or national factors (debt, restructuring of the world economy, negative trends in terms of trade, increasing inequality in the distribution of national wealth, the breakdown of bureaucracies under corruption, to name only a few). Households mobilise the same resources against these hazards as they do against those of the natural environment; having a diversity of livelihood options available, and the skills to adapt flexibly, are just as important. It is equally obvious that rectifying such inequalities or injustices should form a part of any strategy to give rural people in the Sahel the opportunity to help themselves.

We reassert here the thesis that, on the basis of the evidence we have analysed, the 'Sahelian crisis' of degradation can be contained, and that in doing so, the resources of rural communities themselves will play a large part. Pre-eminent among these resources is their labour which, in drawing on a wealth of indigenous technical experience and the best of introduced practices, can create, through an incremental and 'indigenous' intensification of agriculture, more sustainable production systems. Gradualist rather than transformational expectations should therefore underpin the policies of governments and donors, policies which need to be founded both on an improved understanding of the diversity and the dynamics of primary production systems, and on a recognition of the need for unimpeded economic integration between the Sahel and West Africa as a whole.

After decades of 'development', much has changed in the Sahel. Rural households face manifold threats, and their production systems are built on and around the exigencies of environmental and economic risk. And yet sustainability is not an impossible goal about which rural smallholders need to be taught, and in pursuit of which their energies and ideas must be transformed by outside agents. Sustainable livelihoods, and the sustainable environmental management on which they depend, were the stock in trade of Sahelian smallholders

201

long before the 'development decades' of the twentieth century. The search for environmental sustainability, and for economic growth, are the current obsessions of the international development community. That search must begin with the complex, dynamic and tightly managed worlds of Sahelian smallholders themselves. Its success will be revealed in their living experiences in the decades to come.

NOTES

2 DIVERSITY, FLEXIBILITY AND ADAPTABILITY

1 We use the term 'farming' to mean the production of crops, and the term 'livestock production' to mean the keeping and breeding of animals; both are included in the general term 'agriculture' which, however, excludes hunting, fishing, gathering and woodcutting.

2 It may surprise the reader that we do not identify technology as a constraint, in view of the fact that the outside world considers this to be a major limitation to Sahelian productivity. However, from the smallholder's perspective, technologies such as inorganic fertilisers and ploughs *are* available, but often not affordable owing to the capital constraint. Complete 'green revolution' packages for enhancing the yield of food crops have yet to be convincingly demonstrated for the Sahel, so in this sense there *is* a technology constraint.

3 The term 'intensification' is often used without clarifying whether increased inputs of labour, or of capital per hectare are envisaged. We use the first sense in this study, except where stated otherwise. It is an important characteristic of indigenous or labour intensification that labour can create capital (Boserup, 1965; and see Chapter 6, this volume).

4 Greater detail can be achieved by recording labour use by the hour, at the cost of (1) higher levels of intrusion – such data cannot be reliably recalled, and require constant monitoring by research assistants in fields and homes; (2) many more research assistants, and (3) lowering the level of data collection from more participatory to more mechanical. We rejected this option.

4 NEGOTIATING THE RAIN

1 The model uses a wind factor in estimating potential evapotranspiration (Page, 1994: 23). Reducing the wind factor, for example, by windbreaks on farmland, may partly compensate for the increase in temperature.

2 Automatic tipping bucket rain gauges were used, although blockages by airborne material caused two single-season failures (Futchimiram in 1992 and Tumbau in 1994), and faults caused another (Dagaceri in 1996).

3 The use of Nguru daily rainfall data for 1996 in place of that of Dagaceri was considered. Although satisfactory correlations were obtained for monthly rainfall in the years 1992–1995, inspection of the daily data revealed divergent patterns (Buba, 1997). For the purpose of this study, therefore, this option was rejected.

4 The fickle timing of the start of the planting season, and the rapidity of farmers' response, present problems for researchers as well as farmers. In several years we

were unable fully to monitor the planting peak, as information about the first rain either reached us too late for the research team to travel to the village, or it fell just after the team had left following weeks of fruitless waiting.

5 The percentage of available labour used has been found to mirror quite precisely the curves of total labour used during the season; and it makes labour curves of differing magnitude compatible between years or between villages.

6 Migrant hired labour is not available in significant quantities in any of our villages.

6 MAKING THE LAND WORK HARDER

1 The relationship, of course, only addresses *farming* intensity, and assumes a region incompletely converted from natural vegetation to farmland, and where no structural impediments exist to people's access to land. Other productive enterprises, more especially livestock production, also intensify their use of land, but botanical or pedological indicators, rather than changing land use boundaries, are necessary to quantify change.

2 We assume, perhaps riskily, that the areas whose land use has been mapped are representative of their respective districts or, conversely, that the census data for the districts are representative for our study areas.

3 The ratios Tumbau: Dagaceri: Futchimiram: Kaska are 1.0: 0.16: 0.03: 0.04 for the district population density, and 1.0: 0.14: 0.05: 0.05 for agricultural labour available per productive hectare.

4 The possibility exists, but is difficult to verify, that in an absence of unclaimed land, over-sowing is a strategy for maintaining claims to land in excess of a household's ability to cultivate in the year in question.

5 We do not analyse planting because (1) that labour peak is shortlived, (2) our data are incomplete in some villages and some years, owing to the difficulty of anticipating planting rains and the practice of dry planting, and (3) planting may be extended until late in the rainy season.

9 UNDERSTANDING INEQUALITY

1 All names used in this account are fictionalised.

2 Data on livestock ownership were considered too sensitive to ask for in Futchimiram.

3 Owing to the number and small size of irrigated plots, it was not possible in the time available to carry out a survey.

REFERENCES

Abrahams, R.C. and M.M. Kano, 1949 *Dictionary of the Hausa Language.* London: Crown Agents for the Government of Nigeria.

Adams, W.M., 1990 *Green Development: environment and sustainability in the Third World.* London: Routledge.

Adams, W.M., 1991 'Large scale irrigation in northern Nigeria: performance and ideology', *Transactions, Institute of British Geographers,* N.S., 16: 287–300.

Adams, W.M., 1992 *Wasting the rain: rivers, people and planning in Africa.* London/ Minneapolis: Earthscan/Minnesota University Press.

Adams, W.M. and M.J. Mortimore, 1997 'Agricultural intensification and flexibility in the Nigerian Sahel', *Geographical Journal,* 163: 150–160.

Agnew, C.T., 1982 'Water availability and the development of rainfed agriculture in S.W. Niger', *Transactions of the Institute of British Geographers,* 7: 419–457.

Agnew, C.T., 1989 'Sahel drought: meteorological or agricultural?', *International Journal of Climatology,* 9: 371–382.

Alhassan, A.B., 1996 *Soils, water and farming systems in an oasis of the Manga Grasslands,* Ph.D. thesis, Cranfield University.

Amerena, P.M.J., 1982 *Farmers' participation in the cash economy: a case study of two settlements in the Kano Close-Settled Zone of Nigeria,* Ph.D thesis, University of London.

Amin, Samir (ed.), 1974 *Modern migrations in West Africa.* Oxford: Oxford University Press, for the International African Institute.

Andrae, G. and B. Beckman, 1985 *The wheat trap.* London: Zed Books.

Apeldoorn, J. van, 1981 *Perspectives on drought and famine in Nigeria.* London: Allen & Unwin.

Baba, J.M., 1974 *Induced agricultural change in a densely populated district: a study of the existing agricultural system of Kura District and the projected impact of the Kano River Irrigation Project, Kano State, Nigeria,* Ph.D. thesis, Ahmadu Bello University, Zaria.

Bade, J., H. Hengsdijk, G. Kruseman, R. Ruben and P. Roebeling, 1997 'Farm household modelling in a regional setting: the case of Cercle de Koutiala, Mali'. *DLV Report No. 6,* Wageningen Agricultural University.

Barral, H., 1977 'Tiogo, Haute Volta'. *Atlas des structures agraires au sud du Sahara,* 2. Paris: ORSTOM/Maison des Sciences de l'Homme.

Barrow, C.J., 1987 *Water resources and agricultural development in the tropics.* London: Longman.

Barth, H., 1857 *Travels and discoveries in North and Central Africa*, Volume 3. London: Longman.

Behnke, R.H. and I. Scoones, 1991 *Rethinking range ecology: implications for range management in Africa*. London: Overseas Development Institute/International Institute of Environment and Development.

Behnke, R.H., I. Scoones and C. Kerven (eds), 1993 *Range ecology at disequilibrium. New models of natural variability and pastoral adaptation in African savannas*. London: Overseas Development Institute.

Bennett, J.H. (ed.), 1972 'The land resources of north-east Nigeria'. *Land Resources Study 9*. Tolworth: Land Resources Division, Overseas Development Administration.

Bernus, E., 1981 *Touaregs nigériens. Unité culturelle et diversité régionale d'un peuple pasteux*. Paris: ORSTOM.

Berry, S., 1989 *No condition is permanent: the social dynamics of agrarian change in sub-Saharan Africa*. Madison: University of Wisconsin Press.

Blacker, J.G.C., 1993 'Trends in demographic change'. *Transactions of the Royal Society of Tropical Medicine and Hygiene*, 87, Supplement 1: 3–8.

Blaikie, P. and H.C. Brookfield, 1987 *Land degradation and society*. London: Methuen.

Bojö, J. and N. Chee, 1997 'Sahel Operational Review. Second status report'. *Discussion Paper No. 2*, Environment Group. Washington, DC: The World Bank.

Bosc, P.-M., P. Calkins and J.-M. Yung, 1990 *Développement et recherche agricole dans les pays sahéliens et soudaniens d'Afrique*. Montpellier: CIRAD.

Bosc, P.-M., V. Dollé, P. Garia and J.-M. Yung (eds), 1992 'Le développement agricole au Sahel' (5 vols). *Documents Systèmes Agraires No. 17*. Montpellier: CIRAD.

Boserup, E., 1965 *The conditions of agricultural growth: the economics of agrarian change under population pressure*. London: Allen & Unwin.

Boserup, E., 1990 *Economic and demographic relationships in development*. Baltimore, MD: Johns Hopkins University Press.

Bosma, R., K. Bengaly, M. Traoré and A. Roeleveld, 1996 *L'élevage en voie d'intensification. Synthèse de la recherche sur les ruminants dans les exploitations agricoles mixtes au Mali-Sud*. Amsterdam: Institut Royal des Tropiques.

Bourn, D. and W. Wint, 1994 'Livestock, land use and agricultural intensification in sub-Saharan Africa'. *Pastoral Development Network Paper* 37a. London: Overseas Development Institute.

Bradley, P., C. Raynaut and J. Torrealba, 1977 *The Guidimaka region of Mauritania: a critical analysis leading to a development project*. London: War on Want UK.

Bray, M., 1981 *Universal primary education in Nigeria: a study of Kano State*. London: Routledge & Kegan Paul.

Breman, H. and C.T. de Wit, 1983 'Rangeland productivity and exploitation in the Sahel', *Science*, 221: 1341–1347.

Breman, H., 1990 'Integrating crops and livestock in southern Mali: rural development or environmental degradation?' In: Rabbinge, R., J. Goudriaan, H. van Keulen, F.W.T. Penning de Vries and H.H. Laar (eds), *Theoretical production ecology: reflections and prospects*: 277–294. Wageningen: PUDOC.

Brookfield, H.C., 1984 'Intensification revisited', *Pacific Viewpoint*, 25: 15–44.

Brunken, J.N., 1977 'A systematic study of *Pennisetum* sect. *Pennisetum (Gramineae)*', *American Journal of Botany*, 64(2): 161–176.

Buba, L.F., 1992 'Rainfall variability in the Kano region of Nigeria'. Paper presented to the 35th Annual Conference of the Nigerian Geographical Association, Sokoto, 6–10 April 1992.

Buba, L.F., 1997 'African drylands research. Dagaceri/Nguru rainfall analysis', working document (Soils, Cultivars and Livelihoods in North-east Nigeria), Department of Geography, Bayero University, Kano.

Buchanan, K.M. and J.C. Pugh, 1955 *Land and people in Nigeria*. London: London University Press.

Bunting, H., 1992 'Genetic resources and society: preservation of genetic resources'. In: Bureau des Ressources Gènètiques, *Complexes d'espèces, flux de gènes et ressources gènètiques des plantes:* 413–423. Paris.

Bunting, H. and B. Pickersgill, 1996 'What is a plant genetic resource?', *Biologist*, 43 (5): 227–230.

Buntjer, B.J., 1970 'Rural society, the changing structure of *gandu*'. In: Mortimore, M.J. (ed.), 'Zaria and its region. A Nigerian savanna city and its environs': 157–169. *Occasional Paper No. 4.* Department of Geography, Ahmadu Bello University, Zaria.

Busso, C.S., K.M. Devos, M. Mortimore, W. M. Adams, G. Ross, M. J. Ambrose, S. Alldrick and M. D. Gale (forthcoming), 'Genetic diversity among landraces of pearl millet (*Pennisetum glaucum*) under farmer management'.

Cain, M., 1982 'Perspectives on family and fertility in developing countries', *Population Studies*, 36/2: 159–175.

Carney, J. and M. Watts, 1990 'Manufacturing dissent: work, gender and the politics of meaning in a peasant society', *Africa*, 60: 20–22.

Carter, R.C., 1994 'The groundwater hydrology of the Manga Grasslands, northeast Nigeria: importance to agricultural development strategy for the area', *Quarterly Journal of Engineering Geology*, 27: 73–83.

Charney, J.G., 1975 'Dynamics of desert and drought in the Sahel', *Quarterly Journal of the Royal Meteorological Society*, 101: 193–202.

Chattergee, P. and M. Finger, 1994 *The Earth Brokers: power, politics and world development*. London: Routledge.

Chiroma, A.C., 1996 'The farming system of Futchimiram, Yobe State, Nigeria'. *Working Paper No. 4, Soils, Cultivars and Livelihoods in Northeast Nigeria*. Departments of Geography, Cambridge University and Bayero University, Kano.

Cleaver, K.M. and G.A. Schreiber, 1994 *Reversing the spiral. The population, agriculture and environment nexus in sub-Saharan Africa*. Washington, DC: The World Bank.

Cline-Cole, R.A., 1997 'Promoting (anti-)social forestry in northern Nigeria'. *Review of African Political Economy*, 74: 515–536.

Cline-Cole, R.A., H.A.C. Main and J.E. Nichol, 1990 'On fuelwood consumption, population dynamics and deforestation in Africa'. *World Development*, 18 (4): 513–527.

Cline-Cole, R.A., J.A. Falola, H.A.C. Main, M.J. Mortimore, J.E. Nichol and F.D. O'Reilly, 1990 *Wood fuel in Kano*. Tokyo: United Nations University Press.

Clough, P., 1984 'The social relations of grain marketing in Northern Nigeria', *Review of African Political Economy*, 34: 16–34.

Commins, S.K., M.F. Lofchie and R. Payne (eds), 1986 *Africa's agrarian crisis: the roots of famine*. Boulder, CO: Lynne Rienner.

Connah, G., 1988 *African civilisations: precolonial cities and states in tropical Africa, an archaeological perspective*. Cambridge: Cambridge University Press.

Copans, J., 1983 'The Sahelian drought: social sciences and the political economy of underdevelopment'. In: Hewitt, K. (ed.), *Interpretations of calamity from the viewpoint of human ecology:* 83–97. London: Allen & Unwin.

Cour, J.-M., 1994 'Performances du secteur agricole et redistribution de la population en Afrique de l'Ouest'. *Document de Travail no. 12, West Africa Long Term Perspective Study.* Paris: OECD/Club du Sahel.

David, R., 1995 *Changing places? Women, resource management and migration in the Sahel.* London: SOS Sahel UK.

Davies, S., 1996 *Adaptable livelihoods: coping with food insecurity in the Malian Sahel.* Basingstoke/New York: Macmillan/St Martin's Press.

Dalziel, J., 1937 *The useful plants of West Tropical Africa. An Appendix to the flora of West tropical Africa.* London: Crown Agents.

Delgado, C.L., 1979 'The southern Fulani farming system in Upper Volta: a model for the integration of crop and livestock production in the West African savannah'. *African Rural Economy Programme Paper.* East Lansing: Michigan State University.

Dennett, M.D., J. Elston and J.A. Rogers, 1985 'A reappraisal of rainfall trends in the Sahel', *Journal of Climatology,* 5: 353–361.

Dupire, M., 1962 *Peuls nomades. Etude descriptive des Wodaabe du Sahel Nigérien.* Paris: Institut d'Ethnologie.

Ellis, J.E. and D.M. Swift, 1988 'Stability of African pastoral ecosystems: alternate paradigms and implications for development', *Journal of Range Management,* 41: 450–459.

Essiet, E.U., 1994 'Agropastoral adaptation to environmental change in north-east Nigeria. Report on soils sampling and analyses,' Unpublished research paper, Department of Geography, Bayero University, Kano.

Etkin, N. and P. Ross, 1982 'Food as medicine and medicine as food. An adaptive framework for the interpretation of plant utilization among the Hausa of northern Nigeria', *Social Science Medicine,* 16: 1559–1573.

FAO (Higgins, G.M., A.H. Kassam, L. Naiken, G. Fisher and M.M. Shah), 1982 *Potential population-supporting capacities of lands in the developing world. FPA/INT/ 513.* Rome: Food and Agriculture Organisation of the United Nations.

Fairhead, J. and M. Leach, 1996 *Misreading the African landscape: society and ecology in a forest savanna land.* Cambridge: Cambridge University Press.

Ford, R.E., 1998 'Settlement structure and landscape ecology in the Sahel: the case of northern Yatenga, Burkina Faso'. In: Silberfein, M. (ed.), *Rural settlement structure and African development:* 129–166. Boulder, CO: Westview Press.

Forde, C. Daryll, 1946 'The North: the Hausa'. In: Perham, M. (ed.), *The native economies of Nigeria.* London: Faber.

Gadgil, M., F. Berkes and C. Folke, 1993 'Indigenous knowledge for biodiversity conservation', *Ambio,* 22: 151–156.

Gallais, J., 1984 *Hommes du Sahel. Espaces-temps et pouvoirs. Le delta intérieur du Niger 1960–1980.* Paris: Flammarion.

Gastellu, J.M., 1981 'L'égalitarisme économique des Sérèrs au Sénégal'. *Travaux et Documents,* 128. Paris: ORSTOM.

Giri, J. (ed.), 1988 *The Sahel facing the future: increasing dependence or structural transformation.* Paris: OECD.

Glantz, M.H., 1992 'Global warming and environmental change in sub-Saharan Africa', *Global Environmental Change,* 2/3: 183–204.

Goddard, A.D., 1969 'Are Hausa-Fulani family structures breaking up?' *Samaru Agricultural Newsletter*, 11 (3): 34–48. Samaru, Nigeria: Institute for Agricultural Research.

Goddard, A.D., 1973 'Changing family structures among the rural Hausa', *Africa*, 73: 207–218.

Goddard, A.D., J.C. Fine and D.W. Norman, 1967 'A socio-economic study of three villages in the Sokoto Close-Settled Zone. 1, Land and people in a high density rural setting'. *Samaru Miscellaneous Paper No. 34*, Institute for Agricultural Research (Samaru), Ahmadu Bello University, Zaria.

Gould, W.T.S. and M.S. Brown, 1996 'A fertility transition in Sub-Saharan Africa?', *International Journal of Population Geography*, 2: 1–22.

Grégoire, E., 1980 *Etude socio-économique du village de Gourjaie (Département de Maradi, Niger)*. Programme de Recherches sur la Région de Maradi, Université de Bordeaux II.

Grégoire, E. and C. Raynaut, 1980 *Présentation générale du Département du Maradi*. Programme de Recherches sur la Région de Maradi, Université de Bordeaux II.

Grove, A.T., 1958 'The ancient erg of Hausaland and similar formations on the south side of the Sahara', *The Geographical Journal*, 124: 526–533.

Grove, A.T., 1961 'Population densities and agriculture in northern Nigeria'. In: Barbour, K.M. and R.M. Prothero (eds), *Essays on African population:* 115–136. London: Routledge and Kegan Paul.

Grove, A.T., 1976 'A note on the remarkably low rainfall of the Sudan Zone in 1913', *Savanna*, 2: 133–8.

Grove, A.T., 1978 'Geographical introduction to the Sahel', *The Geographical Journal*, 144: 407–415.

Grove, A.T. and A. Warren, 1968 'Quaternary landforms and climate on the south side of the Sahara', *The Geographical Journal*, 134: 194–208.

Haggblade, S., P. Hazell and J. Brown, 1989 'Farm–nonfarm linkages in rural sub-Saharan Africa', *World Development*, 17 (8): 1173–1202.

Hammond, P.B., 1966 *Yatenga. Technology in the culture of a West African kingdom*. New York: The Free Press.

Harlan, J.R., 1995 *The living fields. Our agricultural heritage*. Cambridge: Cambridge University Press.

Harris, F.M.A., 1996 'Intensification of agriculture in semi-arid areas: lessons from the Kano Close-Settled Zone'. *Gatekeeper Series No. 59*. London: Sustainable Agriculture Programme, International Institute for Environment and Development.

Harris, F.M.A., 1998 *Nutrient management strategies of small-holder farmers in a short-fallow farming system in north-east Nigeria. Report to the Natural Resources Systems Programme (Semi-arid Production Systems) of the Department for International Development*. Department of Geography, Cambridge University.

Harris, F.M.A. and B. Bache, 1995 *Nutrient budgets in relation to the sustainability of indigenous farming systems in northern Nigeria. Final Report to the Agronomy and Cropping Systems Programme of the Natural Resources Institute*. Department of Geography, Cambridge University.

Haswell, M.R., 1953 'Economics of agriculture in a savannah village'. *Colonial Research Studies No. 8*, London: HMSO.

Haswell, M.R., 1963 'The changing patterns of economic activity in a Gambia village'. *Overseas Research Publication No. 2*, Department of Technical Cooperation. London: HMSO.

Haswell, M.R., 1975 *The nature of poverty. A case history of the first quarter-century after World War II.* London: Macmillan.

Helleiner, G.K., 1966 *Peasant agriculture, government and economic growth in Nigeria.* London and New Haven, CT: Yale University Press.

Hendy, C.R.C., 1977 'Animal production in Kano State and the requirements for further study in the Kano close-settled zone'. *Land Resource Report* 21. Tolworth: Land Resources Division, Overseas Development Administration.

Hess, T.M., W. Stephens and U.M. Maryah, 1995 'Rainfall trends in the North East Arid Zone of Nigeria 1961–1990', *Agricultural and Forest Meteorology*, 74: 87–97.

Hill, P., 1972 *Rural Hausa: a village and a setting.* Cambridge: Cambridge University Press.

Hill, P., 1977 *Population, prosperity and poverty: rural Kano 1900 and 1970.* Cambridge: Cambridge University Press.

Hill, P., 1986 *Development economics on trial.* Cambridge: Cambridge University Press.

Hiskett, M., 1984 *The development of Islam in West Africa.* London: Longman.

Holling, C.S., 1973 'Resilience and stability of ecological systems', *Annual Review of Ecology and Systematics*, 4: 1–23.

Holling, C.S., 1987 'The resilience of terrestrial ecosystems: local surprise and global change'. In: Clark, W.C. and R.E. Munn (eds), *Sustainable development in the biosphere*: 292–320. Laxenburg, Austria: International Institute for Applied Systems Analysis.

Hollis, G.E., W.M. Adams and M. Aminu-Kano, 1993 *The Hadejia-Nguru Wetlands: environment, economy and sustainable development of a Sahelian floodplain wetland.* Gland, Switzerland: International Union for the Conservation of Nature.

Holmes, J.A., F.A. Street-Perrott, M.J. Allen, P.A. Fothergill, D.D. Harkness, D. Kroon and R.A. Perrott, 1997 'Holocene paleolimnology of Kajemarum Oasis, northern Nigeria: an isotopic study of ostracods, bulk carbonate and organic carbon', *Journal of the Geological Society, London*, 154: 311–319.

Hopen, C.E., 1958 *The pastoral Fulbe family in Gwandu.* London: Oxford University Press for the International African Institute.

Hulme, M., 1987 'Secular changes in wet season structure in Central Sudan', *Journal of Arid Environments*, 13: 31–46.

Hulme, M., 1992 'Rainfall changes in Africa: 1931–1960 to 1961–1990', *International Journal of Climatology*, 12: 685–699.

Hulme, M., 1996 'Climate change within the period of meteorological records'. In: Adams, W.M., A.S. Goudie and A.R. Orme (eds), *The physical geography of Africa*: 88–102. Oxford: Oxford University Press.

Ibrahim, A.M., 1996 'The farming system of Kaska, Yobe State, Nigeria'. *Working Paper No. 3, Soils, Cultivars and Livelihoods in Northeast Nigeria.* Departments of Geography, Cambridge University and Bayero University, Kano.

Iliffe, J., 1995 *Africans: the history of a continent.* Cambridge: Cambridge University Press.

IUCN, 1989 *IUCN Sahel studies, 1989.* Gland, Switzerland: International Union for the Conservation of Nature.

Johnson, V., J. Hill and E. Ivan-Smith, 1997 *Listening to smaller voices: children in an environment of change.* Chard, Somerset: Actionaid.

Jones, M.J. and A. Wild, 1975 *Soils of the West African savanna. The maintenance and improvement of their fertility.* Harpenden: Commonwealth Agricultural Bureau.

Kimmage, K., 1991a 'Small scale irrigation initiatives in Nigeria: the problems of equity and sustainability', *Applied Geography*, 11: 5–20.

Kimmage, K., 1991b 'The evolution of the "wheat trap": the Nigerian wheat boom', *Africa*, 60: 471–500.

Kimmage, K. and W.M. Adams, 1990 'Small-scale farmer-managed irrigation in northern Nigeria', *Geoforum*, 21 (4): 435–443 .

Kimmage, K. and W.M. Adams, 1992 'Wetland agricultural production and river basin development in the Hadejia-Jama'are Valley, Nigeria', *The Geographical Journal*, 158: 1–12.

Koechlin, J., 1980 *Rapport d'étude sur le milieu naturel et les systèmes de production*. Bordeaux: Université de Bordeaux 2.

Koechlin, J., 1997 'Ecological conditions and degradation in the Sahel'. In: Raynaut, C., *Societies and nature in the Sahel*, 12–36. London: Routledge.

Kowal, J.M. and A.H. Kassam, 1978 *Agricultural ecology of savanna. A study of West Africa*. Oxford: The Clarendon Press.

Kowal, J.M. and D.T. Knabe, 1972 *An agroclimatological atlas of the northern states of Nigeria, with explanatory notes*. Samaru: Ahmadu Bello University Press.

Lamb, P.J., 1982 'Persistence of subSaharan drought', *Nature*, 299: 46–47 (2 November).

LRD, 1972 'The land resources of North East Nigeria', *Land Resource Study 9*, Tolworth: Land Resources Division, UK Overseas Development Administration.

Lavigne, P. Delville (ed.), 1998 *Quelles politiques foncières pour l'Afrique rurale? Réconcilier pratiques, légitimité et légalité*. Paris: Karthala-Coopération Francaise.

Leach, M. and R. Mearns, 1996a 'Environmental change and policy: challenging received wisdom in Africa'. In: Leach, M. and R. Mearns (eds), *The lie of the land: challenging received wisdom in African environmental change:* 1–33. London: James Currey/International African Institute.

Leach, M. and R. Mearns, 1996b *The lie of the land: challenging received wisdom in African environmental change*. London: James Currey/International African Institute.

Leeuw, P.N. de, L. Reynolds and B. Roy, 1996 'Nutrient transfers in West African agricultural systems'. In: Powell, J.M., S. Fernandez-Rivera, T.O. Williams and C. Renard (eds), *Livestock and sustainable nutrient cycling in mixed farming systems of sub-Saharan Africa. Vol II, Technical Papers*: 371–392. Addis Ababa: International Livestock Commission for Africa.

Lele, S., 1991 'Sustainable development: a critical review', *World Development*, 19: 607–621.

Lericollais, A., 1972 'Sob. Etude géographique d'un terroir sérèr (Sénégal)'. *Atlas des structures agraires au sud du Sahara, No. 7*. Paris: ORSTOM.

Lericollais, A. (ed.), 1979 *Analyse du changements des systèmes agraires sérèr au Sénégal. Bilan et perspectives des recherches*. Document d'étape. Montpellier: ORSTOM.

Lockwood, J., 1986 'The causes of drought with particular reference to the Sahel', *Progress in Physical Geography*, 10: 111–119.

Lovejoy, P.E., 1986 *Salt of the desert sun*. Cambridge: Cambridge University Press.

Luning, H.A., 1963 *An agro-economic survey in Katsina Province*. Kaduna: Government Printer.

Mabogunje, A.L., 1972 *Regional mobility and resource development in West Africa*. Montreal: McGill-Queen's University Press.

McIntire, J., D. Bourzat and P. Pingali, 1992 *Crop–livestock interactions in sub-Saharan Africa*. Washington, DC: The World Bank.

Macleod, N.H., 1976 'Dust in the Sahel: cause of drought?'. In: Glantz, M. H. (ed.), *The politics of natural disaster: the case of the Sahel drought:* 214–231. New York: Praeger.

McCormick, J.S., 1992 *The global environmental movement: reclaiming paradise.* London: Belhaven.

Mainet, G. and G. Nicolas, 1965 'La vallée du Gulbi de Maradi'. *Etudes Nigériennes No. 16.* Niamey: Centre Nigérien de Recherche en Sciences Humaines.

Marchal, J.-Y., 1983 'Yatenga, nord Haute-Volta. La dynamique d'une espace rural soudanien-sahélien. *Travaux et Documents No. 167.* Paris: ORSTOM.

Meillasoux, C., 1974 'Development of exploitation: is the Sahel famine good business?', *Review of African Political Economy,* 1: 27–33.

Mohammed, S., 1994 'Evaluation of multipurpose uses of trees, shrubs and grasses, conditions of access and smallholder management in a village in the semi-arid North-eastern Nigeria', M.Sc. thesis (Land Resources), Department of Geography, Bayero University, Kano.

Mohammed, S., 1996 'The farming system of Dagaceri, Jigawa State, Nigeria'. *Working Paper No. 2, Soils, Cultivars and Livelihoods in Northeast Nigeria.* Departments of Geography, Cambridge University and Bayero University, Kano.

Morgan, W.B. and J.A. Solarz, 1994 'Agricultural crisis in sub-Saharan Africa: development constraints and policy problems', *The Geographical Journal,* 160: 57–73.

Mortimore, M.J., 1972 'Some aspects of rural–urban relations in Kano, Nigeria'. In: Vennetier, P. (ed.), *La croissance urbaine en Afrique Noire et à Madagascar:* 871–888. Paris: Centre National de la Recherche Scientifique.

Mortimore, M.J., 1982 'Framework for population mobility: the perception of opportunities in Nigeria'. In: Clarke, J.I. and L.A. Kosinski (eds), *Population redistribution in Africa:* 50–57. London: Heinemann.

Mortimore, M.J., 1989 *Adapting to drought: farmers, famines and desertification in West Africa.* Cambridge: Cambridge University Press.

Mortimore, M.J., 1993a 'The intensification of peri-urban agriculture: the Kano Close-Settled Zone, 1964–1986'. In: Turner, B.L. II, G. Hyden and R.W. Kates (eds), *Population growth and agricultural change in Africa:* 356–400. Gainesville: University Press of Florida.

Mortimore, M.J., 1993b 'Northern Nigeria: land transformation under agricultural intensification'. In: Jolly, C.L. and B.B. Torrey (eds), *Population and land use in developing countries: report of a workshop:* 42–69. Washington, DC: National Academy Press.

Mortimore, M.J., 1993c 'The Sahel'. In: Sjöstedt, G. (ed.), *International environmental negotiation.* London: Sage.

Mortimore, M.J., 1995 'Caring for the soil: agricultural expansion, population growth and natural resource degradation in the Sahel'. In: Reenberg, A. and H.S. Marcussen (eds), 'The Sahel: ethnobotany, agricultural and pastoral strategies, development aid strategies': 55–78. *SEREIN Occasional Paper No. 1.* Institute of Geography, University of Copenhagen.

Mortimore, M.J., 1998 *Roots in the African dust. Sustaining the subSaharan drylands.* Cambridge: Cambridge University Press.

Mortimore, M.J. and W.M. Adams (in preparation) 'Farming intensification and its implications for pastoralism in northern Nigeria', paper presented at the University of Giessen, 1997.

Mortimore, M.J. and M. Tiffen, 1995 'Population and environment in time perspective:

the Machakos story'. In Binns, T. (ed.), *People and environment in Africa:* 69–89. Chichester: John Wiley.

Mortimore, M.J. and M. Tiffen, 1996 *Long term change in the West African Drylands: the linkage between rural population density, urbanization, economic change and natural resource management in the Sahel. A preliminary report and research proposal.* London: Overseas Development Institute.

Mortimore, M.J., and B. Turner, 1993 'Crop–livestock farming systems in the semi-arid zone of sub-Saharan Africa. Ordering diversity and understanding change'. *Agricultural Administration (Research and Extension) Network Paper* 46. London: Overseas Development Institute.

Mortimore, M.J. and J. Wilson, 1965 'Land and people in the Kano Close-Settled Zone'. *Occasional Paper No. 1.* Department of Geography, Ahmadu Bello University, Zaria.

Mortimore, M.J., E.U. Essiet and S. Patrick, 1990 *The nature, rate, and effective limits of intensification in the smallholder farming system of the Kano Close-Setled Zone.* Unpublished Report to the Federal Agricultural Coordinating Unit, Ibadan.

Munasinghe, M., 1993 'Environmental issues and decisions in developing countries', *World Development,* 21: 1729–1748.

Nelson, R., 1988 'Dryland management – the desertification problem', *Environment Department Working Paper No. 8.* Washington, DC: The World Bank.

Netting, R. McC., 1968 *Hill farmers of Nigeria. Cultural ecology of the Kofyar of the Jos Plateau.* Seattle: Washington University Press.

Netting, R. McC., 1993 *Smallholders, householders: farm families and the ecology of sustainable intensive agriculture.* Stanford, CA: Stanford University Press.

Nichol, J.E., 1990 'The ecology of wood fuel'. In: Cline-Cole, R.A. *et al.:* 49–70.

Nichol, J.E., 1991 'The extent of desert dunes in northern Nigeria as shown by image enhancement', *The Geographical Journal,* 157: 13–24.

Nicholson, S.E., 1978 'Climatic variations in the Sahel and other African regions during the past five centuries', *Journal of Arid Environments,* 1: 3–24.

Nicholson, S.E., 1996 'Environmental change within the historical period'. In: Adams, W.M., A.S. Goudie and A.R. Orme (eds), *The physical geography of Africa:* 60–87. Oxford: Oxford University Press.

Nicolas, G., 1960 'Un village de la République du Niger: Tassao Haoussa'. Bordeaux: GRID, Université de Bordeaux 2.

Nicolas, G., 1965 *Budgets collectifs de groupements domestiques hausa en République du Niger (Canton de Kantché, Région de Maradi).* Report, Université de Bordeaux 2.

Nicolas, G., 1975 *Dynamique sociale et appréhension du monde au sein d'une société hausa.* Paris: Institut d'Ethnologie.

Niemeijer, D., 1996 'The dynamics of African agricultural history: is it time for a new development paradigm?', *Development and Change,* 27: 87–110.

Norman, D.W., 1967 'An economic study of three villages in Zaria Province: 1. Land and labour relationships. 3, Maps'. *Samaru Miscellaneous Papers Nos 19 and 23.* Zaria: Institute for Agricultural Research.

Norman, D. W., 1972 'An economic study of three villages in Zaria Province: 2. Input–output study' (in two parts). *Samaru Miscellaneous Papers Nos 37 and 38.* Zaria: Institute for Agricultural Research.

Norman, D.W., E.B. Simmons and H.M. Hays, 1982 *Farming systems in the Nigerian savanna. Research strategies for development.* Boulder: Westview Press.

Norman, D.W., J.C. Fine, A.D. Goddard, D.H. Pryor and W.J. Kroeker, 1976 'A socio-economic study of three villages in the Sokoto Close-Settled Zone. 3. Input–output study' (in two parts). *Samaru Miscellaneous Paper* No. 65. Zaria: Institute for Agricultural Research.

OECD, 1988 *The Sahel facing the future. Increasing dependence or structural transformation? Futures study of the Sahel countries, 1985–2010.* Paris: OECD.

Page, J., 1994 'Recent trends in the agroclimatology of Northern Nigeria, with special reference to rainfall changes and their relationship to future predicted climate changes resulting from anthropogenically generated increases in greenhouse gases'. *Report to Cambridge University/Bayero University, Kano Research Project on Agropastoral Adaptation to Intensifying Aridity in North East Nigeria* (internal document). Department of Geography, Cambridge University.

Palmer-Jones, R.W., 1984 'Mismanaging the peasants: some origins of low productivity on irrigation schemes in the North of Nigeria'. In: Adams, W.M. and A.T. Grove (eds), 'Irrigation in Tropical Africa: problems and problem-solving': 96–118. *African Studies Centre Monograph No. 3.* University of Cambridge..

Parker, D.E., C.K. Folland and M.N. Ward, 1987 'Sea surface temperature anomaly patterns and prediction of seasonal rainfall in the Sahel area of Africa'. *Proceedings of the International Geographical Union Study Group on Recent Climate Change*, Sheffield.

Pélissier, P., 1953 'Les paysanes sérères. Essai sur la formation d'un terroir du Sénégal'. *Les Cahiers d'outre-mer*, 6: 105–127.

Penning de Vries, F.W.T. and M.A. Djiteye (eds), 1991 *La productivité des pâturages sahéliens. Une étude des sols, des végétations et de l'exploitation de cette ressource naturelle.* Wageningen: PUDOC.

Pieri, C., 1989 *Fertilité des terres de savanes. Bilan de trente ans de recherche et de développement agricoles au sud du Sahara.* Paris: CIRAD, Ministère de la Co-operation.

Pingali, P., Y. Bigot and H.P. Binswanger, 1987 *Agricultural mechanisation and the evolution of farming systems in sub-Saharan Africa.* Baltimore, MD: The Johns Hopkins University Press.

Piters, B. de Steenhuijsen, 1995 *Diversity of fields and farmers. Explaining yield variations in northern Cameroon.* Doctoral thesis, University of Wageningen.

Powell, J.M., S. Fernandez-Rivera, T.O. Williams and C. Renard (eds), 1995 *Livestock and sustainable nutrient cycling in mixed farming systems of sub-Saharan Africa. Proceedings of an international conference, 22–26 November, 1993. Vol. II: technical papers.* Addis Ababa: International Livestock Centre for Africa.

Pretty, J.N., 1995 *Regenerating agriculture: policies and practice for sustainability and self-reliance.* London: Earthscan.

Prothero, R. Mansell, 1959 *Migrant labour from Sokoto Province, Northern Nigeria.* Kaduna: Government Printer.

Prothero, R. Mansell, 1962 'Some observations on desiccation in North-Western Nigeria', *Erdkunde*, 16: 11–19.

Prothero, R. Mansell (ed.), 1972 *People and land in Africa south of the Sahara.* Oxford: Oxford University Press.

Pullan, R.A., 1974 'Farmed parkland in West Africa', *Savanna*, 3 (2): 119–152.

Rain, D.R., 1997 *Eaters of the dry season*, Ph.D thesis, Department of Geography, Pennsylvania State University.

Raynaut, C., 1973 *Structures normatives et rélations électives: étude d'une communauté villageoise haoussa.* Paris: Mouton.

Raynaut, C., 1977a *Rapport sur les études menées dans l'un des villages témoins: Serkin Haoussa*. University of Bordeaux 2.

Raynaut, C., 1977b 'Lessons of a crisis'. In: Dalby, D., R.J. Harrison-Church and F. Bezzaz (eds), *Drought in Africa 2:* 17–32. London: International African Institute.

Raynaut, C., 1980 *Rapport de synthèse sur les études multidisciplinaires menées dans le Département de Maradi (Niger)*. University of Bordeaux 2.

Raynaut, C., E. Grégoire, P. Janin, J. Koechlin and P. Lavigne Delville, 1997 *Societies and nature in the Sahel*. London: Routledge.

Reardon, T., C. Delgado and P. Matlon, 1992 'Determinants and effects of income diversification among farm households in Burkina Faso', *Journal of Development Studies*, 28 (2): 264–296.

Reenberg, A. and B. Fog, 1995 'The spatial pattern and dynamics of a Sahelian agro-ecosystem', *GeoJournal*, 37.4: 489–499.

Reij, C., I. Scoones and C. Toulmin (eds), 1996 *Sustaining the soil: indigenous soil and water conservation in Africa*. London: Earthscan.

Richards, P., 1985 *Indigenous agricultural revolution: ecology and food production in West Africa*. London: Longman.

Richards, P., 1986 *Coping with hunger: hazard and experiment in an African rice-farming system*. London: Allen & Unwin.

RIM, 1992 *Nigerian livestock resources*. Report to the Federal Department of Livestock and Pest Control Services. Jersey: Resources Inventory and Management.

Rippstein, G. and B. Peyre de Fabregues, 1972 'Modernisation de la Zone Pastorale du Niger' (2 vols). *Étude Agrostologique* 33. Maison Alfort: Institut d'Élévage et de Médecine Vétérinaire de Pays Tropicaux.

Rochette, R.M., 1989 *Le Sahel en lutte contre la désertification. Leçons d'experiences*. Eschborn: Deutsche Gesellschaft für Technische Zusammenerbeit.

Rodney, W., 1972 *How Europe underdeveloped Africa*. London: Bogle-l'Ouverture Publications.

Roe, E., 1991 'Development narratives, or making the best of blueprint development', *World Development*, 19: 287–300.

Rosenzweig, C. and M.L. Parry, 1994 'Potential impacts of climate change on world food supply', *Nature*, 367: 133–138.

Rouch, J., 1956 'Migrations au Ghana', *Journal de la Société des Africanistes,* 26: 33–196.

Roy, E. le, A. Karsenty and A. Bertrand, 1996 *La sécurisation foncière en Afrique. Pour une question viable des ressources renouvelables*. Paris: Karthala.

Sandford, S., 1983 *Management of pastoral development in the Third World*. Chichester: John Wiley.

Sautter, G., 1962 'A propos de quelques terroirs d'Afrique occidentale, essai comparatif'. *Etudes Rurales*, 6: 24–82.

Savonnet, G., 1970 'Pina. Etude d'un terroir de font pionnier en pays dogari (Haute-Volta)'. *Atlas des structures agraires au sud du Sahara, No 4*. Paris: Mouton.

Schultz, T.W., 1964 *Transforming traditional agriculture*. New Haven, CT: Yale University Press.

Scoones, I. (ed.), 1991 *Wetlands in drylands: the agroecology of savanna systems in Africa*. London: International Institute for Environment and Development.

Scoones, I., 1994 *Living with uncertainty: new directions in pastoral development in Africa*. London: Intermediate Technology Publications.

Scoones, I., 1996 *Hazards and opportunities. farming livelihoods in dryland Africa. Lessons from Zimbabwe*. London: Zed Books/International Institute for Environment and Development.

Siddle, D. and K. Swindell, 1990 *Rural change in tropical Africa: from colonies to nation-states*. Oxford: Blackwell.

Sinclair, A.R. and J.M. Fryxell, 1985 'The Sahel of Africa: ecology of a disaster'. *Canadian Journal of Zoology*, 63: 987–994.

Smaling, E., 1993 *An agro-ecological framework for integrated nutrient management with special reference to Kenya*, Doctoral thesis, Wageningen Agricultural University.

Smith, M.G., 1955 'The economy of Hausa communities of Zaria'. *Colonial Office Research Series 16*. London: HMSO.

Snrech, S., with J.-M. Cour, A. de Lattre and J.D. Naudet, 1994 *West Africa long term perspective study. Preparing for the future: a vision of West Africa in the year 2020. Summary report*. Paris: OECD/Club du Sahel.

Somerville, D., 1986 *Drought and aid in the Sahel. A decade of development cooperation*. Boulder, CO: Westview Press.

Stamp, L.D., 1938 'Land utilisation and soil erosion in Nigeria', *The Geographical Review*, 28: 32–45.

Stenning, D.J., 1959 *Savannah nomads. A study of the Wodaabe pastoral Fulani of western Bornu Province Northern Region, Nigeria*. Oxford: Oxford University Press.

Stern, R.D., M.D. Dennett and D.J. Garbutt, 1981 'The start of the rains in West Africa', *Journal of Climatology*, 1: 59–68.

Stone, G.D., 1997 *Settlement ecology: the social and spatial organisation of Kofyar agriculture*. Tucson: University of Arizona Press.

Stone, G.D., R. McC. Netting and M. P. Stone, 1990 'Seasonality, labour scheduling, and agricultural intensification in the Nigerian savanna', *American Anthropologist*, 92 (1): 7–23.

Sutton, J.E.G., 1989 'Towards a history of cultivating the fields', *Azania*, 24: 99–112.

Swift., J., 1996 'Desertification: narratives, winners and losers'. In: Leach and Mearns (eds): 73–90.

Swindell, K. and Iliya, M., 1992 'Accumulation, consolidation and survival: non farm incomes and agrarian change in northwest Nigeria'. *Project on African Agriculture Working Paper Series*, 6. New York: Social Science Research Council.

Thomas, D. S. G. and T. Middleton, 1994 *Desertification: exploding the myth*. Chichester: John Wiley.

Tiffen, M., M. Mortimore and F. Gichuki, 1994 *More people, less erosion: environmental recovery in Kenya*. Chichester: John Wiley.

Toulmin, C., 1992 *Cattle, women and wells. Managing household survival in the Sahel*. Oxford: The Clarendon Press.

Tucker, C.C., J. Compton, H.E. Dregne and W.W. Newcomb, 1991 'Expansion and contraction of the Sahara Desert from 1980 to 1990'. *Science*, 19, July: 229–301.

Turner, B., 1984 'Changing land use patterns in the *fadamas* of northern Nigeria'. In: Scott, E.P., *Life before the drought*: 149–170. London: Allen & Unwin.

Turner, B., 1997 'Land cover change around four villages in north-east Nigeria: methodology and results'. *Working Paper No 5, Soils, Cultivars and Livelihoods in North-east Nigeria*. Departments of Geography, Cambridge University and Bayero University, Kano.

Turner, B.L. II, R.W. Kates and G. Hyden (eds), 1993 *Population growth and agricultural change in Africa*. Gainesville: University Press of Florida.

Udo, R. K., 1975 *Migrant tenant farmers of Nigeria. A geographical study of migrations in Nigeria*. Lagos: African Universities Press.

UNEP, 1992 *World atlas of desertification*. Nairobi: United Nations Environment Programme.

United Nations, 1977 *Desertification: its causes and consequences*. Oxford: Pergamon Press.

Van der Pol, F., 1992 'Soil mining: an unseen contributor to farm income in southern Mali'. *Royal Tropical Institute Bulletin* 325, Royal Tropical Institute, Amsterdam.

Walker, J. and P.R. Rowntree, 1977 'The effect of soil moisture on circulation and rainfall in a tropical model', *Quarterly Journal of the Royal Meteorological Society*, 103: 29–46.

Wallace, T., 1980 'Agricultural projects and land in northern Nigeria', *Review of African Political Economy*, 17: 59–70.

Wallace, T., 1981 'The Kano River Project in Nigeria: the impact of an irrigation scheme on productivity and welfare'. In: Heyer, J., P. Roberts, and G. Williams (eds), *Rural development in tropical Africa*: 281–305. London: Macmillan.

Warren, A., 1996 'Desertification'. In: Adams, W.M., A.S. Goudie and A.R. Orme (eds), *The physical geography of Africa*: 342–355. Oxford: Oxford University Press.

Warren, A. and M. Khogali, 1991 *Assessment of desertification and drought in the Sudano-Sahelian region, 1985–1991*. New York: United Nations Sudano-Sahelian Office.

Watts, M.J., 1983 *Silent violence: food, famine and peasantry in northern Nigeria*. Berkeley: University of California Press.

Watts, M.J., 1984 'The demise of the moral economy: food and famine in the Sudano-Sahelian region in historical perspective'. In: Scott, E.P. (ed.), *Life before the drought*: 228–248. London: Allen & Unwin.

Watts, M.J., 1989 'The agrarian question in Africa: debating the crisis'. *Progress in Human Geography*, 13: 1–14.

Wiggins, S., 1995 'Changes in African farming systems between the mid-1970s and the mid-1980s'. *Journal of International Development*, 7: 807–848.

Williams, G., 1981 'The World Bank and the peasant problem'. In: Heyer, J., P. Roberts and G. Williams (eds), *Rural development in Tropical Africa*: 16–51. London: Macmillan.

Wilson, J.P., G.W. Burton, J.D. Zango and I.O. Dicko, 1990 'Diversity among pearl millet landraces collected in central Burkina Faso', *Crop Science*, 30: 40–43.

World Bank, 1990 *Sub-Saharan Africa: from crisis to sustainable growth: a long-term perspective study*. Washington, DC: The World Bank.

World Bank (D. Spurling), 1995 *Rural women in the Sahel and their access to agricultural extension. Sector study. Overview of five country studies, June, 1995*. Washington, DC: The World Bank, Africa region.

Young, A. and A.C.S. Wright, 1984 'Rest period requirements of tropical and sub-tropical soils under annual crops'. In: FAO, *Report on the second FAO/UNFPA expert consultation on land resources for populations of the future*. Rome: Food and Agriculture Organisation of the United Nations.

Yusuf, M.A., 1994 'The influence of agricultural management on present and potential soil quality under smallholder conditions: a case study of Tumbau (Kano Close-Settled

Zone).' *M.Sc thesis (Land Resources)*, Department of Geography, Bayero University, Kano.

Yusuf, M.A., 1996 'The farming system of Tumbau, Kano State, Nigeria'. *Working Paper No 1, Soils, Cultivars and Livelihoods in North-east Nigeria*. Departments of Geography, Cambridge University and Bayero University, Kano.

INDEX